T0262623

Introduction to the Interstellar Medium

The gas and dust between the stars emit across the electromagnetic spectrum and are found in a range of physical conditions from diffuse plasmas to cold, dense, and molecular. Through their study we see how quantum processes shape the structure of our Galaxy and fluid mechanics sets the stellar mass scale. The interstellar medium is a very broad subject with layers of complexity, a long history, and a steady flow of new results. This comprehensive yet accessible textbook provides a self-contained one-semester course for advanced undergraduate or beginning graduate students. It is written in a style that students can follow by themselves and allows instructors to use class time to go deeper into the details or show applications to current research. It makes extensive use of publicly accessible data to illustrate specific points and to encourage students to learn by performing their own analyses.

JONATHAN P. WILLIAMS is an astronomer at the University of Hawaii at Manoa. His research in the interstellar medium has ranged from the formation of the giant molecular clouds that form stars to the evolution of disks around young stars that give birth to planets. He has written pedagogical reviews on these topics and taught a wide variety of courses at the undergraduate and graduate level. This textbook builds on his course notes that have been widely used and class-tested over almost two decades.

Introduction to the Interstellar Medium

Jonathan P. Williams
University of Hawaii, Manoa

CAMBRIDGE
UNIVERSITY PRESS

CAMBRIDGE
UNIVERSITY PRESS

University Printing House, Cambridge CB2 8BS, United Kingdom

One Liberty Plaza, 20th Floor, New York, NY 10006, USA

477 Williamstown Road, Port Melbourne, VIC 3207, Australia

314–321, 3rd Floor, Plot 3, Splendor Forum, Jasola District Centre, New Delhi – 110025, India

79 Anson Road, #06–04/06, Singapore 079906

Cambridge University Press is part of the University of Cambridge.

It furthers the University's mission by disseminating knowledge in the pursuit of education, learning, and research at the highest international levels of excellence.

www.cambridge.org
Information on this title: www.cambridge.org/9781108480802
DOI: 10.1017/9781108691178

First published 2021

Printed in the United Kingdom by TJ Books Limited, Padstow Cornwall

A catalogue record for this publication is available from the British Library.

ISBN 978-1-108-48080-2 Hardback

Additional resources for this publication at www.cambridge.org/interstellarmedium

Contents

Preface

A basic understanding of the Interstellar Medium (ISM) should be a core part of an astrophysics curriculum. The gas and dust between the stars emit across the electromagnetic spectrum and the ISM is found in a range of physical conditions from diffuse plasma to cold, dense, and molecular. Through its study we see how quantum processes shape the structure of our Galaxy and fluid mechanics sets the stellar mass scale. There is a tremendous richness here and plenty to keep theorists, observers, and instrumentalists busy. Although most astronomers will spend their careers in different areas, every telescope looks through the ISM and many analyses must account for its effect on the data or use it as a diagnostic tool.

The study of the ISM is a relatively mature field with a wide variety of phenomena that are generally well understood. Of course, there remains much to be discovered and a continual flow of interesting papers. However, it is easy for a new student to get lost in the details of recent work before fully grasping the fundamentals. This pedagogical book is intended to introduce general concepts for an undergraduate senior, beginning graduate student, or anyone with a solid physics background that is interested. It does not describe the latest findings in a fast-paced research world though I hope that its broader perspective might help more seasoned researchers make new connections.

The philosophy of this book is that doing is the best way of learning. The internet has become the great leveler and everyone can now access all sorts of wonderful data. This is a boon to our abstract field. As an encouragement for students to explore these goldmines for themselves, almost all the figures in this book were created from scratch using publicly available datasets acquired through open archives, and python notebooks to recreate them are available at interstellarmedium.github.io. The figures are black and white to keep costs down but beautiful color images of all these phenomena are available at your fingertips. In the same spirit, questions at the end of each chapter walk through additional steps of derivations in the text or recreate plots. The length is appropriate for a one semester course and I have tried to write at a level that a physics major can follow by themselves so that an instructor can use lectures to augment the material with additional details or recent findings. To limit distractions, I have not given references in the main text but include the most pertinent at the end of each chapter, and make suggestions for other books and review articles there. At a more advanced level, there is at least one paper per day, and often many more, on arXiv (arxiv.org/archive/astro-ph) that apply concepts from this book in current research.

This book grew out of lecture notes that I wrote over almost two decades of graduate teaching at the University of Hawaii, with their genesis in the notes I took as a student myself in an ISM class co-taught by the "dream team" of Chris McKee and Carl Heiles at the University of California at Berkeley 30 years ago. I am grateful to them for nurturing my interest in this field and also to influential teachers further back in my academic journey including John Green and Doug Gough at the University of Cambridge and beginning with Jeff Aspinall at Aylesbury Grammar School. This is my way to pay it forward.

Getting to this point has been a long, but enjoyable, journey and would not have been possible without a lot of help. First and foremost, the Institute for Astronomy provided the environment to get me started and the freedom to finish this project. I have been fortunate to have worked with and learned from many talented faculty, postdocs, and students on many aspects of the ISM. Their names are on our papers but I would like to single out Leo Blitz who introduced me to the Rosette molecular cloud which features prominently throughout these pages. Cathie Clarke and Chris DuPree bravely read the first draft and gave invaluable comments. The aforementioned Chris McKee continued to educate me through his detailed suggestions.

Last and definitely not least, I would like to thank my family, Laura, Nicholas, and Julian, for their love and support throughout.

Chapter 1
Introduction

1.1 What Is the Interstellar Medium?

The Interstellar Medium is the stuff between the stars. It is made up almost entirely of gas with a small smattering of tiny particles called dust grains. The ISM is readily apparent to the naked eye in the dark smudges in the Milky Way. With small telescopes or binoculars, fuzzy nebulae can be seen. First photographic plates and, nowadays, electronic imagers add up the light over long exposure times and reveal the colors of these features including the reddening of starlight, blue reflection nebulae, and a rainbow of emission lines. Vast clouds of atomic hydrogen were discovered with radio telescopes in the 1950s and then molecular clouds in the next two decades. The advent of space astronomy opened the Universe to exploration across the electromagnetic spectrum and the ISM showed its presence at every wavelength, including hot gas in X-rays, absorption lines in the ultraviolet, and warm dust around young stars in the infrared.

The different wavelengths reveal different components of the ISM. Figure 1.1 shows the Milky Way Galaxy at optical, infrared, and radio wavelengths to highlight the main components that we will discuss in this book. At top, in the optical ($\lambda \approx 0.5\,\mu\text{m}$) we mainly see starlight punctuated by dark clouds. By studying how the light is blocked, we learn about the interstellar dust in these clouds. In the next panel down, the near-infrared ($\lambda \approx 2\,\mu\text{m}$) images shows the stellar distribution more clearly, including the bulge near the Galactic Center. The fact that the dark clouds are now almost transparent tells us how big the dust grains are. Energy is conserved so if dust absorbs starlight, it must also radiate. The third panel down shows the emission from dust at far-infrared wavelengths ($\lambda \approx 350\,\mu\text{m}$). We discuss dust in Chapter 4. Now moving

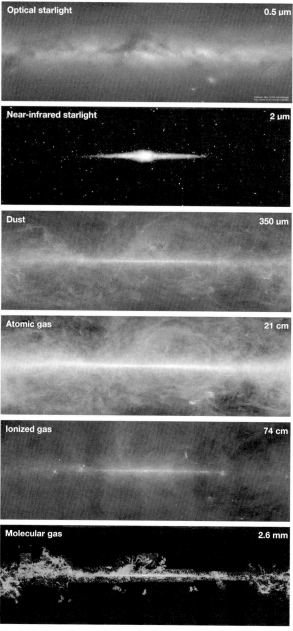

Fig. 1.1. The Galaxy at different wavelengths showing different components of the ISM. The horizontal axis is Galactic Longitude varying from $-180°$ to $+180°$ and the vertical axis is Galactic Latitude varying from $-60°$ to $+60°$.

to the gas, the fourth panel shows the emission from atomic hydrogen in the $\lambda = 21$ cm spectral line. There is a striking similarity to the dust emission, which shows that the gas and dust are well mixed in the ISM and, as we show in Chapter 5, the ratio between them is constant. The gas is concentrated along the Galactic plane but it also extends well above, indicating the ISM is a very dynamic place. These fluid motions are the subject of Chapter 8. The longest wavelength image, $\lambda = 74$ cm, in the fifth panel shows radiation from ionized gas. The emission is from accelerating charged particles in low-density plasma. The motions may be either relatively slow, such as the thermal motions in bubbles of warm gas around hot stars, but energetic phenomena such as supernovae can shock particles to relativistic speeds, producing strong emission as they gyrate around magnetic field lines. The way in which the intensity varies with wavelength distinguishes each case. We discuss ionized regions in Chapter 6. The relativistic particles, known as cosmic rays, permeate the ISM and play an important role in its heating and chemistry. Finally, the bottom panel shows the distribution of molecular gas in the Galaxy, as detected through the $\lambda = 2.6$ mm spectral line of carbon monoxide. The physics and chemistry of molecular clouds are discussed in Chapter 7. These regions are the coldest, densest parts of the ISM and the places where stars are born, the subject of Chapter 9. Having focused on the different phases of the ISM in each chapter, we return to a Galactic perspective and consider the ecology of the ISM in Chapter 10. All these components are of course found in other galaxies and some in the space between galaxies. Indeed, the two features in the HI map that stand out just right of center and below the plane are the Large and Small Magellanic Clouds, and a keen eye can spot the Andromeda galaxy on the left. The book concludes with an extragalactic perspective in Chapter 11.

The dust and gas, hot and cold, ionized, atomic, or molecular, are the subjects of this book. Most of the Universe is actually dark matter and energy and, in a sense, these are part of the ISM too. However, these components do not interact with baryonic matter or radiation and they are so broadly distributed that they have no significant net gravitational force on the scale of the objects that we consider here. Thus, in all respects, dark matter and energy have no effect on the processes described in this book. Apart from a perhaps self-serving remark that any astrophysical search for them almost certainly requires a thorough understanding of the radiation from the ISM, they are not discussed further.

1.2 The Vacuum of Space

By terrestrial standards, the ISM is an almost perfect vacuum. The typical distance between stars is about $2\,\mathrm{pc} = 6 \times 10^{16}\,\mathrm{m}$, about 100 million times greater than the radius of our Sun and 4000 times greater than the size of its heliosphere. The ISM fills the vast space in between but its total mass in the entire Galaxy is only $M \approx 7 \times 10^9\,M_\odot$. Approximating the volume as a cylinder with radius $R \approx 10^4\,\mathrm{pc}$, scale height $H \approx 250\,\mathrm{pc}$, this implies an average density $\rho = M/2\pi R H \approx 3 \times 10^{-21}\,\mathrm{kg\,m^{-3}}$. Hydrogen accounts for about 74% of the mass, helium for about 25%, and other elements make up the remaining 1%. The corresponding hydrogen number density is, therefore, $n_\mathrm{H} \approx 10^6\,\mathrm{m^{-3}}$. This is easier to remember as about one particle per cubic centimeter.[1]

In comparison, we can calculate the density of air from the ideal gas law using the pressure at sea level, $P = 10^5\,\mathrm{N\,m^{-2}}$ (1 bar), and temperature, $T \approx 300\,\mathrm{K}$, to get $n = P/kT \approx 2 \times 10^{25}\,\mathrm{m^{-3}}$, a full 19 orders of magnitude higher than the average density in the ISM. To help visualize just how enormous a difference this is, consider a cylinder between you and a wall in the room you are in, about 10 m away. Now imagine that cylinder stretched lengthwise, extending from the edge of the Solar System to the center of the Galaxy, 8.2 kpc away. Each cylinder would have the same number of atoms!

There is a small overlap between the highest vacuums we can achieve in laboratories with the densest regions that we discuss in this book, $n \sim 10^{12}\,\mathrm{m^{-3}}$, but in general the extremely low densities in space mean that particle collisions are relatively rare, which allows us to observe some physical processes that we don't see on Earth.

For particles with a cross-section σ, moving at speed v, simple geometric considerations illustrated in Figure 1.2 show that the timescale between collisions is $t_\mathrm{coll} = (n\sigma v)^{-1}$. The motions in the air are predominantly thermal with a characteristic speed $= (kT/m)^{1/2} = 300\,\mathrm{m\,s^{-1}}$ at room temperature. The most common particle is molecular nitrogen with $\sigma \simeq 4 \times 10^{-19}\,\mathrm{m^2}$, which gives $t_\mathrm{coll} \sim 4 \times 10^{-9}\,\mathrm{s}$. This is so short that if a collision excites an internal energy level of a particle, it will often de-excite through a second collision rather than through spontaneous emission of radiation. The distributions of kinetic and internal energies of the gas are therefore in collisional equilibrium. We will cover these concepts in more detail in the following chapter.

---

[1] Much of the astronomy literature still uses the cgs system though the recommended standard by the International Astronomical Union, and the approach I will take in this book, is to use SI units. The Appendix gives values for commonly used constants in both systems.

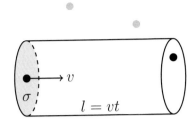

Fig. 1.2. A particle with cross-section σ moving at speed v sweeps out a volume $\sigma v t$ in time t. If it collides with one particle during this time, then the density of particles $n = 1/(\sigma v t)$. Therefore the collision timescale is $t_{coll} = 1/(n\sigma v)$ and the mean free path $l = v t_{coll} = 1/(n\sigma)$.

In the ISM, the typical particle is hydrogen with a smaller cross-section, $\sigma \simeq 6 \times 10^{-21}\,\mathrm{m}^{-2}$, slightly higher speed, $v \simeq 3000\,\mathrm{m\,s}^{-1}$, but much lower density, $n \simeq 10^{6}\,\mathrm{m}^{-3}$, implying $t_{coll} \sim 2 \times 10^{10}\,\mathrm{s} \simeq 10^{3}$ years. This is so long that collisional excitation is often followed by radiative de-excitation. This affects the energy level distribution of gas particles and means that nebulae are lit with spectral lines that we don't see in terrestrial environments, and that temperatures are set not by conduction but by radiation.

Due to the low densities, the chemistry in molecular clouds is very different from the terrestrial setting. Simultaneous three-body collisions are almost non-existent and many important astrochemical reactions, including the formation of H_2, occur on dust grain surfaces.

The mean free path, $l = (n\sigma)^{-1}$, is of course much smaller in the relatively dense Earth's atmosphere, $\sim 0.1\,\mu\mathrm{m}$, compared to the ISM, $\sim 10^{14}\,\mathrm{m}$. However, the relevant comparison is to the size scales of the objects under consideration, for example a human, building, or mountain on Earth and an interstellar cloud in the ISM. In each case, gas particles will collide many times during their passage across an object of interest. We can therefore treat the ISM as an astrophysical fluid when studying how a gas cloud moves in response to, for example, external pressure forces or its own gravity.

1.3 Why Study the ISM?

The ISM is a unique physical and chemical laboratory on account of its ultra-low densities. It is also where stars are born and it is enriched by newly produced elements when stars die. It can outshine direct starlight in distant galaxies and it is where we can see the first atoms in the Universe.

However, the ISM is also messy, irregularly shaped, turbulent, magnetized, constantly shocked, and irradiated on scales from stellar to galactic. It is a complex subject and the details can sometimes overwhelm. Nevertheless, the ISM is also beautiful, both in the literal sense, as in images of colorful nebulae, and in the physics that helps us understand our origins and the way the Universe works.

The ISM, almost by definition, is everywhere and it affects all sorts of observations, sometimes as a nuisance through the absorption of starlight perhaps, but more often as an essential complement for understanding the Galaxy. In particular, it is the place where we can learn about the birth and death of stars, and the chemical enrichment and evolution of galaxies. Although we only provide a basic introduction to each of these subjects in this short book, the goal is for the reader to gain some physical insights into the big picture that underlies many areas of current research.

Chapter 2
Observations

Astronomy is an observationally driven science in the sense that theories generally try to explain what we see rather than the other way around. The HI 21 cm line is a notable exception, but most of the phenomena that we discuss in this book were first discovered at a telescope and only then understood through physical calculations.

The temperatures in the ISM range from less than ten to over a million kelvin. These, and other non-thermal processes, produce a range of radiation processes that affect light from radio to X-rays. As Figure 1.1 shows, different wavelengths show a different aspect of the ISM. Here we give a brief primer on observational nomenclature and techniques across the electromagnetic spectrum.

2.1 Radiation Diagnostics

Light can be detected over a broad range of wavelengths or frequencies, or in a narrow spectral feature, termed **continuum** and **line** respectively (Figure 2.1). In general, the former results from macroscopic dust particles or plasma where there is a broad range of energy states, whereas the latter is due to quantized jumps between the internal energy levels of atoms, ions, or molecules in diffuse gas.

The variation of continuum radiation over a wide (orders of magnitude) range of frequency is termed the **spectral energy distribution** and abbreviated to SED. Many of the emission or absorption lines in the ISM are Gaussian in shape and can be characterized by a central frequency, peak intensity, and width. However, line profiles can be more complex with, for example, multiple peaks or extended emission in the wings away from the line center.

7

Fig. 2.1. A schematic plot of intensity versus wavelength showing continuum emission as a large-scale, slowly varying term and spectral line features, single peaked emission and absorption, and a double-peaked line profile at the right.

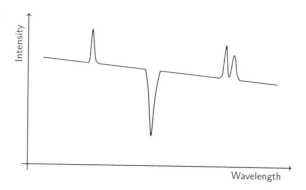

Light may also be polarized, and this is an important diagnostic in the ISM. Examples include scattering and emission by dust and effects induced by magnetic fields. We will discuss the relevant details as we encounter them in later chapters.

All images have a finite resolution, or intrinsic limit to distinguish two nearby objects. For a telescope with a (circular) aperture of diameter D, the **angular resolution** is

$$\theta = \frac{\lambda}{D} \text{ radians.} \tag{2.1}$$

Often this limit is quoted as 1.22θ, which is actually the scale at which the intensity is one-half its peak value. In practice, the scale at which two objects can be distinguished depends on their brightness profile and the quality (or signal-to-noise ratio) of the data. In almost all cases, the angle is small and the corresponding linear scale is $L = \theta d$ where d is the distance to the object.

Spectrometers are similarly limited in their ability to distinguish two peaks in line profiles, expressed in terms of wavelength as $\Delta\lambda$. The **spectral resolution** is defined to be

$$R = \frac{\lambda}{\Delta\lambda}, \tag{2.2}$$

and its value depends on the instrument design. For sources that are moving at speed v projected along our line of sight, the **Doppler effect** shifts the wavelength by a fractional amount $\Delta\lambda/\lambda = v/c$ where c is the speed of light. The spectral resolution therefore converts to a velocity resolution, c/R.

2.2 Telescopes across the Electromagnetic Spectrum

The Universe emits at all wavelengths but the Earth's atmosphere only lets a small range of these pass through to the ground. The ionized upper skin of the atmosphere reflects decameter and longer wavelength

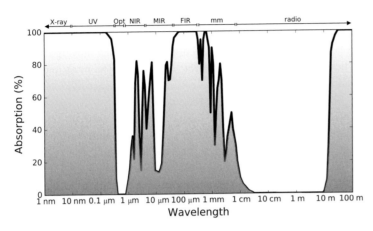

Fig. 2.2. Absorption of the atmosphere over 11 orders of magnitude in wavelength, from X-rays to the radio.

radio waves back into space. Lower down in the denser, molecular part of the atmosphere, ultraviolet radiation is scattered and absorbed predominantly by nitrogen and ozone respectively. Still lower, infrared radiation is absorbed mainly by water and carbon dioxide. Nevertheless, there are small regions where the atmosphere is transparent, the optical window from $\lambda \approx 0.4$ to $0.7\,\mu m$, where our eyes (not coincidentally) work, and the radio window from $\lambda \approx 1\,cm$ to $10\,m$, which was very important for early studies of the ISM and remains so today. In between these two windows, there are narrow regions where the atmosphere partially transmits light. Because the transmission is greater at high and dry sites where the water vapor is low, most observatories are built on mountaintops.

The absorption of the atmosphere across the electromagnetic spectrum from a good astronomical site is shown in Figure 2.2. To observe wavelengths where the absorption is high requires either high-altitude balloon-borne telescopes or space observatories. The detection techniques also vary greatly across such a wide wavelength range. These considerations affect our observational view of the ISM and are discussed for each wavelength regime below.

Radio

Radio observations are at the long-wavelength end of the electromagnetic spectrum, $\lambda > 300\,\mu m$ ($\nu < 1\,THz$). The upper end has a practical limit of about $10\,m$ set by the ionosphere. As the energy of a photon at these long wavelengths is very small, $E = h\nu = hc/\lambda \simeq 10^{-26}-10^{-21}\,J$ where h is the Planck constant, the light behaves more as a wave than a particle. Detection is therefore made via antennas and amplification of the current induced by the wave. Radio waves can be collected and focused through parabolic dishes as with optical telescopes.

Despite the low energy of radio waves, the detected continuum radiation comes from energetic phenomena, generally ionized gas that has been heated either by hot stars or contains rapidly moving charged particles in a strong magnetic field (Chapter 6). The most important spectral line is from atomic hydrogen at 21 cm (Chapter 5).

Because the wavelengths are long, radio telescopes have low angular resolution. The largest steerable radio dish is the Green Bank Telescope and has a diameter of 100 m. At its shortest operating wavelength of 1 cm, the angular resolution $\theta \simeq 10^{-4}$ radians $\simeq 20''$. This is fine for looking at many regions of extended, low surface brightness features in the ISM but limits our ability to study smaller objects such as star-forming cores in the Galaxy or clouds in external galaxies. Pairs of telescopes can be linked together to measure the interference pattern of the sky, similarly to Young's double-slit experiment, with an effective resolution $\theta = \lambda/B$ where B is the distance, termed baseline, between the two telescopes. The baselines can be much greater than any single-dish telescope with correspondingly higher resolution (lower θ). Multiple telescopes can be linked together to form an **interferometric array**. Because the phase information of the astronomical signal is preserved in radio wave detection, this is easier to perform than in the infrared or optical where only the energy of the photons is measured. The Very Large Array consists of 27 antennas, each 25 m in diameter, that can be moved into configurations ranging up to a maximum baseline of 36 km, corresponding to arcsecond resolution at 21 cm. Very long baseline interferometry (VLBI) links telescopes across the Earth with baselines of several thousands of kilometers and a resolution less than a milli-arcsecond. VLBI observations of quasars, which are fixed point sources, sets the coordinate reference system for the sky and is so accurate that continental drift can be measured to a precision of millimeters per year.

At shorter wavelengths, $\lambda < 1$ cm, atmospheric absorption by water attenuates the astronomical signal but observations can be carried out at very dry sites. Millimeter-wavelength astronomy reveals the thermal emission from cool dust grains (Chapter 4) and rotational transitions of molecules (Chapter 7). To amplify these high-frequency ($\nu > 30\,\mathrm{GHz}$) waves, the incoming astronomical source is mixed with a synthesized source and down-converted to a few GHz where longer wavelength radio signal processing techniques can be used, a technique known as **heterodyne detection**. The Atacama Large Millimeter Array, located on a high plateau in Chile at an altitude of 5000 m, consists of 50 dishes of 12 m diameter that can be arranged in compact configurations with baselines $B < 160$ m, to very extended configurations with a maximum baseline of 16 km. At $\lambda = 1$ mm, the range of achievable resolution is

$\sim 0\overset{''}{.}015 - 1\overset{''}{.}5$. It has unprecedented ability for studying star and planet formation in nearby regions and the molecular ISM in distant galaxies.

The digital revolution that has affected all of society has greatly benefited radio astronomy. The detected signal can be analyzed over very small frequency intervals through auto-correlation techniques to provide high spectral resolution, $R > 10^6$, equivalent to a velocity resolution $< 0.1 \, \text{km s}^{-1}$. The future of radio astronomy lies in ever larger arrays of antennas with data processing by ever faster computers.

Infrared

The infrared region lies between the optical and radio, $\lambda \sim 0.8 - 300 \, \mu\text{m}$, and correspondingly reveals stars and many components of the ISM. From Wien's law, a blackbody with temperature T peaks at $\lambda = 3000 \, \mu\text{m}/T$, which gives an indication of the temperatures of objects that emit in this region, $T \sim 10 - 4000 \, \text{K}$.

This wide range is split into three sub-regions. At the shortest wavelengths, the near-infrared goes up to $\sim 2.4 \, \mu\text{m}$, beyond which the atmospheric transparency sharply decreases. Starlight dominates the emission in this range but vibrational molecular lines can be detected in hot or shocked regions (Chapter 7), as well as important diagnostics of nebulae, for example hydrogen recombination lines (Chapter 6). Then follows the mid-infrared, which extends to $\sim 25 \, \mu\text{m}$ and reveals warm dust and molecular lines. This is also called the thermal infrared as terrestrial objects emit strongly at these wavelengths. The far-infrared begins at $25 \, \mu\text{m}$, where the atmosphere is almost completely opaque. The ISM emits strongly at these wavelengths as there are copious amounts of dust with temperatures of a few tens of kelvin (Chapter 4). Star-forming regions and starburst galaxies strongly emit throughout this whole range, and their evolutionary state can be traced by following the peak of emission from the far-infrared to the near-infrared (Chapters 10 and 9).

At the shorter wavelength range, in the near- and mid-infrared, the light is detected in arrays of **photodiodes**. Essentially, the incoming photons eject electrons in a semi-conductor that cross over a band gap and create an electric current. In the far-infrared, the light is absorbed and the consequent increase in temperature is measured, a technique known as **bolometry**.

The atmosphere is fairly transparent in the near-infrared and telescopes and instruments are similar to those used in the optical (see below). The mid-infrared is more difficult to observe from the ground both because the atmosphere is more opaque and also because the telescope and instruments themselves produce thermal radiation that generally dwarfs the astronomical signal. The far-infrared can only be observed from aircraft, balloons, or space. Many space-based infrared observatories have made important breakthroughs in all areas of astrophysics. The first all-sky maps were created by the Infrared Astronomical Satellite (IRAS) in the 1980s. These low-resolution (few arcminutes) maps have served as guides for many subsequent missions, most notably the Spitzer and Herschel satellites, which have imaged large regions of the Galaxy at much higher sensitivity and resolution down to a few arcseconds.

Most spectrometers in the infrared rely on prisms or gratings to separate different wavelengths. The spectral resolution is lower than can be achieved through the heterodyne technique in the radio but can reach $R \sim 10^5$ in the near-infrared, which allows kinematic studies of hot gas. More typically $R \sim 10^2 - 10^3$ which is sufficient to differentiate between various molecular and atomic lines and broad features from dust grains.

Optical

The optical, or visual, region is the narrowest band of the electromagnetic spectrum, spanning $\lambda \sim 0.4-0.7 \, \mu$m and defined by the range over which the human eye can detect light. Most astronomical telescopes work in the optical (and often beyond into the near-infrared) regime, and observations can be carried out anywhere on a cloudless night. The angular resolution is theoretically small, $\lambda/D \sim 1''$, for a 1 m telescope but is limited in practice for larger apertures due to turbulence in the atmosphere. Telescopes with diameters $\gtrsim 2$ m are therefore located on sites that provide stable atmosphere and high image quality ("good seeing"). The ultimate mitigation for atmospheric effects is to go above it, and achieving the highest resolution was the primary motivation for the Hubble Space Telescope. However, in the ~ 30 years since its launch, adaptive optics technology has improved to the point that ground-based telescopes can now correct much of the atmospheric distortion to achieve comparable or better resolution than Hubble, at least over small fields of view.

Optical photons have energies of a few electronvolts and can readily eject electrons from semi-conductors. **Charge-coupled devices** (CCDs) read out large arrays of such photodetectors to provide digital imaging capability. The digital revolution that has given us ubiquitous megapixel cameras on cheap phones is also behind gigapixel arrays on large telescopes that map degree-scale fields in a single shot.

Large-scale optical imaging benefits ISM studies through maps of dust extinction and reddening (Chapter 4). Optical spectroscopy reveals nebulae emission lines that diagnose the physical properties of ionized regions (Chapter 6). As with the infrared, high spectral resolution, $R \sim 10^5$, is achievable (and particularly important for Doppler detection of exoplanets for example) but most observations of the ISM are made at more moderate resolution, $R \sim 10^3$, as the broader bandwidth captures more photons.

High Energy

The atmosphere is completely opaque to high-energy photons with wavelengths $\lambda \lesssim 0.3 \, \mu m = 300 \, nm$, and observations at these wavelengths are carried out almost exclusively from space.

Ultraviolet ($\lambda \sim 10-300 \, nm$) telescopes and instruments are similar in design to optical. Massive stars with spectral type O produce strong continuum emission over this range and can ionize hydrogen at 91.2 nm (Chapter 6). This region also contains the Lyman α line at 121.6 nm, which is an essential diagnostic of the intergalactic medium (Chapter 11). Absorption line spectroscopy of other elements reveals the composition of interstellar gas (Chapter 5).

At the shortest wavelengths, $\lambda \lesssim 10 \, nm$, extending beyond the left side of Figure 2.2, X-rays and gamma-rays locate the most extreme physical environments: neutron star surfaces and black hole accretion disks, also stellar flares and, most relevant for this book, the hot ionized medium with temperatures $\sim 10^6 \, K$ (Chapters 8 and 11). X-rays pass through most materials but can be reflected at low incident angles. Telescopes bring the light to a focus through nested groups of mirrors arranged in a cone along the line of sight. The radiation is detected through CCDs that share similarities with their optical counterparts but count individual photons, or bolometers that measure the increase in temperature caused by the absorption of a photon. The literature generally uses photon energy, $E = h\nu = hc/\lambda$, ranging from keV to MeV, rather than wavelength or frequency in this regime.

2.3 The Virtual Observatory

The sky has been surveyed across the electromagnetic spectrum and the maps are publicly accessible in a manner that is not too different from looking up locations on the Earth using Google maps. Indeed such open-access data were used to create Figure 1.1. A good starting point to learn about the data available on an astronomical object is the Simbad database at simbad.u-strasbg.fr/simbad, which provides images from all-sky data through the Aladin sky atlas. Skyview provides FITS format image files for user-specified regions from X-ray to radio (skyview.gsfc.nasa.gov).

In addition, almost all research observatories archive the data that they collect. These archives are generally online, searchable, and open-access after a proprietary period of typically one year. One example for NASA space telescopes is mast.stsci.edu. It is also becoming more common for research papers to make available the data and catalogs that they present and often the analysis routines that they used.

The resources available from these "open skies" policies democratize astronomy and allow people (and countries) to carry out meaningful research without their own expensive infrastructure investments. Many of the figures in this book were consciously created using public data as a demonstration of their applicability to the concepts described here and to provide starting points for readers who want to delve into the data by themselves.

Notes

This is a very brief introduction to observational techniques and intended only to provide basic familiarity and some terminology that we use later in the book. Most research telescopes and space-based missions have good webpages that describe their optical design and instrumentation. There are also many textbooks that go into much greater detail than here. Examples include the all-purpose *An Introduction to Modern Astrophysics* by Carroll and Ostlie (2006) and the more specific *Measuring the Universe* by Rieke (2017).

Chapter 3
Essential Background Physics

The ISM is a low-density mixture of particles that interact with each other through collisions and radiation. Starlight and shocks heat and ionize the gas. The ISM cools when internally generated photons escape. In the coldest regions, atoms combine into molecules. Although the topics that we cover in this book use fundamental physical principles that date back to the turn of the previous century, their application allows us to understand modern astrophysical phenomena.

3.1 Statistical Mechanics

When we observe spectral lines from the ISM, we are detecting photons that have been emitted from quantized states in individual gas particles. We will variously consider spin–orbit coupling of an electron and a proton, electron orbital levels, and molecular vibrational and rotational states. The equations of statistical mechanics allow us to connect these measurements to macroscopic properties of the gas, such as temperature and density, which we can then use to study its physical nature. This is because collisions distribute the internal energy of the system between random motions of particles relative to each other and the energy within individual particles.

The range of speeds of gas particles with mass m at temperature T is described by the **Maxwell–Boltzmann distribution**,

$$f(v) = \sqrt{\frac{2}{\pi} \left(\frac{m}{kT}\right)^3} \, v^2 e^{-mv^2/2kT}, \tag{3.1}$$

where $f(v)dv$ is the fraction of particles with speeds between v and $v + dv$ and k is the Boltzmann constant. Figure 3.1 plots the distribution for hydrogen gas at $T = 10, 100$ and 1000 K. The first two

Fig. 3.1. The
Maxwell–Boltzmann
distribution for three
different temperatures,
showing the range of
thermal speeds in different
ISM environments.

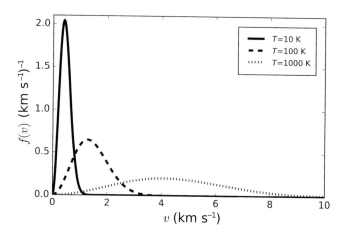

correspond, approximately, to the temperatures of cold molecular and
atomic gas respectively. Speeds are a few kilometers per second in
these regions but we will also encounter warmer atomic and ionized
gas where temperatures are several thousand kelvin and speeds are of
order $10 \, \mathrm{km \, s^{-1}}$. The peak of the distribution, where $df/dv = 0$,
implies that the most probable speed is $(2kT/m)^{1/2}$. The mean square
value, $\langle v^2 \rangle = \int v^2 f \, dv = (3kT/m)^{1/2}$, implies a mean kinetic energy,
$\frac{1}{2}m\langle v^2 \rangle = \frac{3}{2}kT$, per particle.

Collisions between particles in the gas redistribute energy among
them and also in the quantized energy states of the particles themselves.
In **thermodynamic equilibrium**, the proportion of particles in different
levels follows a **Boltzmann distribution**,

$$\frac{n_2}{n_1} = \frac{g_2}{g_1} e^{-(E_2 - E_1)/kT}, \tag{3.2}$$

where n_i and E_i are the number densities and energies of each level,
$i = 1, 2$, and T is the same temperature used to describe the motions
in the gas; g_i is the statistical weight, also known as degeneracy, of
a level and represents different quantum states that have the same
energy. Mathematically, it can be thought of as the number of different
eigenvectors with the same eigenvalue in a solution to the Schrödinger
equation. Formulae will be given as we come across different cases.

The distribution of energy levels is also affected by radiation.
Collisionally excited particles may decay through spontaneous emission. Photons can also excite particles to higher states or induce other
excited particles to emit. We describe these processes further in the
chapter. In thermodynamic equilibrium the resulting radiation field is
described by the **Planck function**,

$$B_\nu(T) = \frac{2h\nu^3}{c^2} \frac{1}{e^{h\nu/kT} - 1}, \tag{3.3}$$

where ν is the frequency, h is the Planck constant, and c is the speed of light. This is also known as **blackbody radiation** as it represents a situation of multiple interactions between photons and matter as occurs in a perfectly absorbing and perfectly emitting "blackbody".

The ISM, as a whole, is not in thermodynamic equilibrium because collisional rates are low and radiation fields can be very strong close to stars or other energy sources. However, there are cases where the system under study can be approximated as being in **local thermodynamic equilibrium** (LTE). A common example is when densities are sufficiently high for collisional rates to exceed spontaneous emission.

When LTE does not apply, we can still describe the population levels of a particle using the Boltzmann form in Equation 3.2, by defining an **excitation temperature**,

$$T_{ex} = \frac{E_2 - E_1}{k \, \ln(g_2 \, n_1 / g_1 \, n_2)}. \tag{3.4}$$

This has units of K but is not a physical temperature and may vary from one level pair to another. It can even be negative if populations are inverted due to radiative pumping of upper states. The difference between kinetic and excitation temperatures parameterizes how far the distribution of states is from LTE.

Similarly, the radiation field in a system out of LTE may not be a blackbody. To see what form the radiation field can take, we need to look at the interaction between light and matter in the ISM.

3.2 Radiative Transfer

When we observe an astronomical source we are detecting the energy that it radiates at a specific frequency, ν. The amount of energy, E_ν (where the subscript indicates that is a function of freqeuncy), depends on the frequency range, $\Delta\nu$, length of time, t, light collecting area, A, and the solid angle, Ω = area of source/(distance)^2, over which the emission is measured. Over a small enough range of each of these quantities such that the radiation field is constant, the dependence becomes linear and can be factored out. We therefore define the **specific intensity**, I_ν, through the differential form,

$$dE_\nu = I_\nu \, dA \, d\Omega \, d\nu \, dt. \tag{3.5}$$

I_ν has the rather abstract units of $\text{W m}^{-2} \, \text{Hz}^{-1}$ steradian^{-1} and its use is best shown through example. Its importance is that it is intrinsic to the source itself and not the measurement. By deriving I_ν from data, we can therefore learn about the process by which the source emits and derive properties such as temperature or mass. Conversely, we can construct

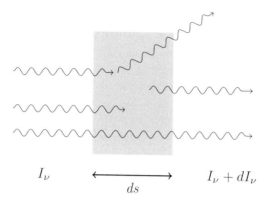

Fig. 3.2. Schematic of the losses through scattering and absorption and gains through emission as radiation passes through a material, changing the specific intensity by dI_ν over pathlength ds.

$I_\nu \quad\longleftrightarrow\quad I_\nu + dI_\nu$

ds

models of a source to predict how much radiation we would detect in various observational scenarios.

We now consider radiation within the source itself. Figure 3.2 illustrates how I_ν is modified through gains (emission) and losses (absorption and scattering). Over a small pathlength, ds, we can write

$$dI_\nu = j_\nu ds - \kappa_\nu I_\nu ds, \tag{3.6}$$

where j_ν is the **emission coefficient** and κ_ν is the **absorption coefficient**. The first term describes the spontaneous emission of photons whereas the second term scales with I_ν because the number of photons lost is proportional to the number that pass through the material.

Radiation is affected not by the distance it travels but by its interaction with matter. This motivates a change from pathlength to **optical depth**,

$$d\tau_\nu \equiv \kappa_\nu ds. \tag{3.7}$$

This leads to the **equation of radiative transfer**,

$$\frac{dI_\nu}{d\tau_\nu} = S_\nu - I_\nu, \tag{3.8}$$

where

$$S_\nu \equiv \frac{j_\nu}{\kappa_\nu} \tag{3.9}$$

is the **source function**. This has the general solution

$$I_\nu(\tau_\nu) = I_\nu(0)e^{-\tau_\nu} + \int_0^{\tau_\nu} S_\nu\, e^{(\tau_\nu - \tau')}\, d\tau', \tag{3.10}$$

which states that the specific intensity is the sum of an attenuated background source and material emission coefficient modified by absorption along the line of sight.

For matter in thermodynamic equilibrium at a single temperature T, the specific intensity is a blackbody everywhere (Equation 3.3),

$I_\nu = B_\nu(T)$, and $dI_\nu/d\tau_\nu = 0$. This implies $S_\nu = B_\nu$ and therefore $j_\nu = \kappa_\nu B_\nu$. However, since the emission and absorption coefficients are quantum properties of the matter and independent of the macroscopic environment, this condition must hold generally with the excitation temperature T_{ex} that describes the level populations,

$$j_\nu = \kappa_\nu B_\nu(T_{ex}). \tag{3.11}$$

This relation between how well a material emits and absorbs is known as **Kirchoff's law** of thermal radiation. The reasoning used here that allows us to extend a relationship between quantities derived in a particular set of circumstances to a more general case is because, in equilibrium, every elementary process is statistically balanced by its exact reverse process. This is known as the principle of **detailed balance** and we will see its application relating collisional rates later.

The equation of radiative transfer has a simple solution if the excitation temperature is constant along the line of sight,

$$I_\nu = I_\nu(0)e^{-\tau_\nu} + B_\nu(T_{ex})(1 - e^{-\tau_\nu}). \tag{3.12}$$

In many cases, however, the radiation affects the physical state of the matter so the excitation temperature is not constant. Iterative or Monte Carlo (solving for many photon paths probabilistically) approaches are then used to solve the equation.

3.3 Quantized Absorption and Emission

The discrete energy levels of a quantum particle produce **spectral lines**. Emission or absorption occurs only for photons with energies $h\nu$ close to the energy difference, ΔE, between the levels. The lines occur over a range of frequencies due to the Heisenberg uncertainty principle which relates the energy and lifetime of the excited state. For particles in isolation, the lifetime is the inverse of the spontaneous emission rate, and the resulting natural broadening of the line is very small. At very high densities, the lifetime is determined by the collisional rate and is termed pressure broadening. This is unimportant in the low-density ISM, but can be dominant in stellar or planetary atmospheres, including Earth's.

The strongest effect in most ISM lines is the Doppler effect which shifts the frequency by a fractional amount $\Delta\nu/\nu = v/c$ for line-of-sight speed v. This affects both the line center due to bulk motion of the gas and broadens the line due to internal motions. The **line profile**, $\phi(\nu)$, parameterizes the proportion of photons emitted or absorbed by a transition at each frequency and is normalized such that $\int_0^\infty \phi(\nu)\,d\nu = 1$.

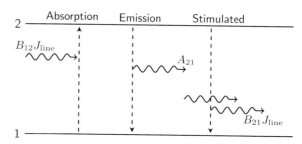

There are three radiative transitions between a pair of levels, with rates characterized by the Einstein A and B coefficients (Figure 3.3). A particle in the high-energy state, 2, may spontaneously decay to the lower state, 1, through emission of a photon at a rate A_{21}. A particle in the lower state may absorb a photon at a rate $B_{12} J_{\text{line}}$, where J_{line} is the specific intensity integrated over the line profile and averaged over direction,

$$J_{\text{line}} = \frac{1}{4\pi} \int \int I_\nu \, \phi(\nu) \, d\nu \, d\Omega. \tag{3.13}$$

In addition, and a critical insight from Einstein, is that the radiation field can also stimulate the production of photons from particles in the upper state at a rate $B_{21} J_{\text{line}}$.

In thermodynamic equilibrium, the absorption rate must be balanced by the spontaneous and stimulated emission rates,

$$n_1 \, B_{12} \, J_{\text{line}} = n_2 (A_{21} + B_{21} \, J_{\text{line}}). \tag{3.14}$$

In the case that the population ratio is defined by the Boltzmann equation and the specific intensity by a blackbody, this implies

$$g_1 \, B_{12} = g_2 \, B_{21}, \quad B_{21} = \frac{c^2}{2h\nu^3} A_{21}. \tag{3.15}$$

The derivation is left as an exercise for the reader in question 2 at the end of this chapter.

The principle of detailed balance discussed above means that this relationship between quantum properties of the matter must hold generally, regardless of the radiation field or level populations. With a little more algebra, we can also extend the same reasoning to a multi-level system and find the same relationship between the Einstein coefficients for any given transition.

The frequency dependence $A_{21}/B_{21} \propto \nu^3$ implies that spontaneous emission is much more likely than stimulated emission for widely separated transitions. Nevertheless, stimulated emission can be important in astrophysical contexts at low frequencies, as we will see regarding the atomic hydrogen 21 cm line in Chapter 5.

If we now consider the group of particles as a whole, photons are spontaneously emitted, with energy $h\nu$ distributed with a line profile $\phi(\nu)$, at a volumetric rate $n_2 A_{21}$ over 4π steradians, where n_2 is the number density in the upper state. The emission coefficient is, therefore,

$$j_\nu = \frac{h\nu}{4\pi} n_2 A_{21}\phi(\nu). \tag{3.16}$$

Similarly, we can relate the absorption coefficient to the Einstein B coefficients,

$$\kappa_\nu = \frac{h\nu}{4\pi}(n_1 B_{12} - n_2 B_{21})\phi(\nu), \tag{3.17}$$

where the stimulated emission term, B_{21}, acts as negative absorption. Using Equations 3.2 and 3.15, we can reduce this to

$$\kappa_\nu = \frac{h\nu}{4\pi}\left(1 - e^{-h\nu/kT_{ex}}\right) n_1 B_{12}\phi(\nu). \tag{3.18}$$

Note that if $n_2 B_{21} > n_1 B_{12}$ then $\kappa_\nu < 0$ and the material amplifies the specific intensity (Equation 3.6) because the stimulated emission exceeds absorption. This can indeed happen in particular cases where an upper energy level is "pumped" leading to population inversion, $n_2/g_2 > n_1/g_1$ (Equation 3.15), and is the process behind the ubiquitous lasers in everyday life. It was first demonstrated in centimeter wavelength transitions of ammonia and subsequently observed in the ISM. Such masers (microwave amplification of stimulated emission of radiation) are commonly found in star-forming regions where there is dense molecular gas accompanied by a strong radiation field.

3.4 Flux Density and Luminosity

We need to relate the radiation emitted by the source to the radiation measured by an observer. Consider a telescope imaging an object at distance d with an angular resolution θ. This means that the instrument can receive light from the source over a solid angle $\Omega_s \propto \theta^2$. Figure 3.4 draws the paths that arrive at a particular point in the telescope and ultimately the detector. The total amount of radiation received is the sum

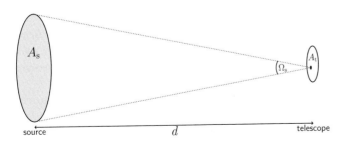

Fig. 3.4. The path of light rays from source emission to detection from the perspective of a point in the collecting area of the telescope.

of all these light rays over the collecting area of the telescope, A_t. After correcting for atmospheric and instrumental effects, we measure the total energy per unit time per unit frequency, received from the source, $P_\nu^{rec} = I_\nu^{rec} \Omega_s A_t$, where I_ν^{rec} is the specific intensity received by the observer.

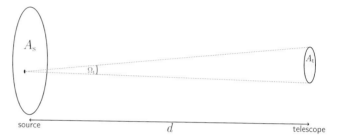

Fig. 3.5. The path of light rays from source emission to detection from the perspective of a point on the surface of the source that lies within the telescope field-of-view.

Figure 3.5 draws the paths from a particular point on the source that lie within a solid angle Ω_t and can reach the telescope. The total amount of radiation emitted toward the telescope is the sum of all these light rays over the area A_s. The total energy per unit time per frequency is therefore $P_\nu^{emit} = I_\nu^{emit} \Omega_t A_s$, where I_ν^{emit} is the specific intensity emitted by the source.

The two descriptions describe, in alternate but equivalent ways, all the possible light paths from the source to the telescope. The received and emitted powers must therefore be the same, $P_\nu^{rec} = P_\nu^{emit}$, which implies

$$I_\nu^{rec} = I_\nu^{emit} \frac{\Omega_t A_s}{\Omega_s A_t} = I_\nu^{emit} \frac{A_t/d^2}{A_t} \frac{A_s}{A_s/d^2} = I_\nu^{emit}. \tag{3.19}$$

Thus, we receive the same specific intensity at the telescope as is emitted at the source.

Think of specific intensity as a surface brightness. A heuristic way to visualize the above result is to imagine (or try!) looking at a uniformly lit wall and walking toward it. As you get closer, a field-of-view with fixed angular size will see a progressively smaller region of the wall, but this is exactly balanced by the inverse square law describing the spreading of the light rays from the wall. Consequently the brightness remains the same. The situation is different for a point source (e.g., a light bulb) as all the light comes from a region smaller than your field-of-view so you do not receive less light from a smaller area as you walk closer but you still gain from the inverse square law. The bulb therefore becomes brighter as you walk toward it.

In practice, observations have a finite resolution so we do not measure the specific intensity directly but rather its integral over an angular area, known as the **flux density**,

$$F_\nu = \int_{\Omega_s} I_\nu \, d\Omega. \qquad (3.20)$$

The SI values for astrophysical objects are very small so this is generally expressed in units of the **jansky** (Jy) where $1\,\mathrm{Jy} = 10^{-26}\,\mathrm{W\,m^{-2}\,Hz^{-1}}$. The **flux** is the integral of the flux density over frequency,

$$F = \int F_\nu \, d\nu, \qquad (3.21)$$

with units of $\mathrm{W\,m^{-2}}$.

To determine the energy per unit time, we integrate flux over area. Most sources emit isotropically so, using the constancy of I_ν and considering a sphere centered on the source with radius d, the **monochromatic luminosity** is

$$L_\nu = 4\pi \, d^2 \, F_\nu, \qquad (3.22)$$

and the **bolometric luminosity** is

$$L_{\mathrm{bol}} = \int L_\nu \, d\nu. \qquad (3.23)$$

We have defined quantities in terms of frequency but we could equally have considered the wavelength version of the specific intensity, I_λ. Given that the energy must be the same over the same bandwidth, $\Delta\lambda = c/\nu^2 \, \Delta\nu$ (defined to be positive), this implies $\lambda \, I_\lambda = \nu \, I_\nu$ and, therefore,

$$\lambda \, F_\lambda = \nu \, F_\nu. \qquad (3.24)$$

A source SED is often plotted as this quantity versus frequency or wavelength. Shown in logarithmic axes, we can look at areas under the curve and, because $\int \nu F_\nu d\ln\nu = \int F_\nu d\nu$, immediately see where the source emits the bulk of its energy. As we will see, an object's SED provides essential clues to diagnose its physical nature.

3.5 The Hydrogen Atom

Hydrogen is the most common element in the Universe. It is ten times more abundant than helium and ten thousand times more than carbon, nitrogen, and oxygen. Most other elements are many orders of magnitude even less common. Consequently, hydrogen is the dominant constituent of the ISM, both by number and mass, and the way in which it interacts with radiation is key to understanding the different components of the ISM. We therefore quickly review some salient features of the hydrogen atom here.

The ionization potential of hydrogen is $E_{\mathrm{IP}} = 13.6\,\mathrm{eV}$, which is equivalent to the energy of far-ultraviolet photons, $\lambda = hc/E_{\mathrm{IP}} = 91.2\,\mathrm{nm}$,

Fig. 3.6. The Bohr model
for the electronic energy
levels in a hydrogen atom.
Ultraviolet photons can
ionize the atom and
recombination produces
numerous spectral lines over
a wide range of energies.

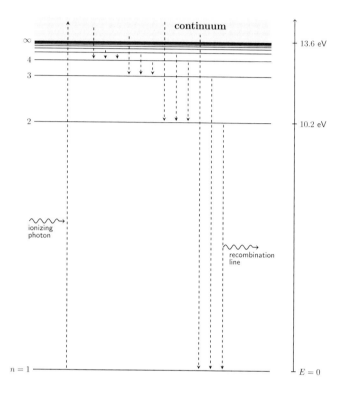

or high-speed collisions, $v = (2E_{\mathrm{IP}}/m)^{1/2} \sim 50\,\mathrm{km\ s^{-1}}$. A neutral atom has bound levels labeled by quantum number, $n = 1, 2, 3 \ldots$, which are well described by the Bohr model,

$$E_n = E_{\mathrm{IP}} \left(1 - \frac{1}{n^2} \right), \qquad (3.25)$$

illustrated in Figure 3.6.

The most striking feature is the large spacing between the lower two levels, $n = 1, 2$. The energy difference, $\Delta E = E_2 - E_1 = 10.2\,\mathrm{eV}$, is much greater than the average level found in all but the hottest parts of the ISM, $T > \Delta E/k = 1.2 \times 10^5$ K. Collisional excitation in a neutral hydrogen cloud is therefore rare and almost all the atoms will be in the ground state. Consequently, an isolated hydrogen cloud in the ISM will neither absorb nor emit optical or infrared photons as these have energies of a few tenths to a few eV, which is below the $n = 2$ level. However, the ground state, $n = 1$, is split by coupling between the electron and proton spin resulting in hyperfine structure of $6\,\mu$eV corresponding to $\lambda = 21$ cm.

In ionized gas, an electron that recombines with a proton may form a hydrogen atom in an excited state, $n > 1$. The electron will then cascade down the energy ladder, producing a series of recombination lines, $n_2 \rightarrow n_1$. These can vary from small energy jumps, producing radio or infrared emission at large n_1, to the near-infrared for $n_1 = 3$ (Paschen series), optical for $n_2 = 2$ (Balmer series), and ultraviolet for $n_1 = 1$ (Lyman series). These concepts and more on the nomenclature are elaborated upon as we encounter them in subsequent chapters.

Notes

This is a cursory treatment of several fundamental concepts that are used throughout the book. Many introductory physics courses and textbooks can provide more background, if required. As with the previous chapter, a suitable broad astronomical reference is *An Introduction to Modern Astrophysics* by Carroll and Ostlie (2006).

Questions

1. From the Heisenberg uncertainty principle, heuristically show that the natural broadening of a line is $\Delta v \approx A/2\pi$, where A is the spontaneous emission rate. Compare its value with thermal broadening for the Hα line at 10^4 K.

2. Derive the relationship between the Einstein A and B coefficients in Equation 3.15. (Hint: consider a very narrow spectral line such that $\phi(v)$ is a delta-function and generalize.) Show that the emission and absorption coefficients for a spectral line satisfy Kirchoff's law.

3. The specific intensity of a star is, to first order, a blackbody. For a given effective temperature, T_{eff}, and stellar radius, R, derive its bolometric luminosity. Look up values for these parameters and calculate this formula for the Sun. Plot the SED ($v F_v$ versus v) for $T_{\text{eff}} = 10^3, 5 \times 10^3, 10^4$ K.

4. What is the sign of the excitation temperature for a maser line?

Chapter 4
Dust

Most of the ISM is gas but a small fraction, about 1% by mass, is in solids. Interstellar dust is composed of elements and compounds that condense out of expanding and cooling stellar winds from giant stars and supernovae. Unlike the gas, which mainly emits and absorbs light in narrow spectral lines, dust particles interact strongly with radiation over a very broad wavelength range. About 30% of the total radiation emitted by the Galaxy is from dust and this fraction can be much higher for galaxies that are undergoing a burst of star formation. Because of its outsized influence on starlight, dust is the starting point for our exploration of the ISM.

4.1 Extinction and Reddening

Interstellar dust can be observed with the naked eye (from a dark site), as dark patches against the diffuse background starlight of the Galaxy. The small dust particles, or **grains**, absorb and scatter light making the background stars appear fainter. With enough dust, essentially all the optical light is blocked and we see a **dark cloud**.

Imaging from blue to red shows that dust absorption and scattering is wavelength dependent. Figure 4.1 shows that the dark cloud Barnard 68 is most prominent in the optical blue band but barely perceptible in the infrared beyond 2 μm. This dramatic difference over a factor of about 5 in wavelength tells us something about both the size of the dust grains and the amount of dust along the line of sight.

Another way to think about this effect is that stars not only become fainter but also redder. We quantify this through the **extinction**, $A(\lambda)$, in the magnitude system,

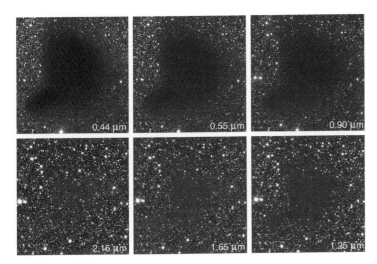

Fig. 4.1. Images of the Barnard 68 dark cloud at six different wavelengths, labeled in the lower right corner of each panel and progressively increasing clockwise from the top left. Credit: ESO.

$$m(\lambda) = M(\lambda) + 5\log_{10}\left(\frac{d}{10\,\text{pc}}\right) + A(\lambda), \qquad (4.1)$$

where $m(\lambda), M(\lambda)$ are the apparent and absolute magnitudes of a star at distance d. We define the **color** as the relative difference in magnitude between two wavelengths,

$$E(\lambda_1 - \lambda_2) = m(\lambda_1) - m(\lambda_2)$$
$$= M(\lambda_1) - M(\lambda_2) + A(\lambda_1) - A(\lambda_2), \qquad (4.2)$$

which is independent of distance. Now consider observations of two stars, $i = 1, 2$, of the same spectral type and therefore same absolute magnitude at each wavelength. If one of the spectral twins is nearby, it should have negligible extinction and the difference in colors between the two stars is

$$E_1(\lambda_1 - \lambda_2) - E_2(\lambda_1 - \lambda_2) = A_1(\lambda_1) - A_1(\lambda_2). \qquad (4.3)$$

We can therefore determine the wavelength dependence of extinction, known as the **extinction curve**.

The shape of the extinction curve differs slightly from one sightline to another but in a systematic manner across wavelengths such that it can be described by a single parameter. This is most often characterized by the ratio of the extinction in the visual band, V at 0.551 μm, to the color relative to the blue band, B at 0.445 μm,

$$R_V = \frac{A_V}{E(B-V)}. \qquad (4.4)$$

R_V is known as the **selective extinction** and has a mean value of 3.3 with a dispersion of 0.2 in the Galaxy. It depends only on the dust and is independent of the distance to the source or the source properties. As the extinction becomes smaller at longer wavelengths (Figure 4.1), the effect of different R_V is most important in the optical and ultraviolet.

It is illustrative to convert the extinction from magnitudes, a logarithmic scale where 5 magnitudes corresponds to a factor of 100 in brightness, to the ratio of the measured flux, F_ν, with its intrinsic value, F_ν^0,

$$A(\nu) = -2.5 \, \log_{10} \left(\frac{F_\nu}{F_\nu^0} \right). \tag{4.5}$$

However, using Equation 3.10 with no source term, we can also write this in terms of optical depth, $F_\nu = F_\nu^0 \, e^{-\tau_\nu}$, and therefore

$$A(\nu) = 2.5 \, \log_{10} e^{\tau_\nu} = 1.086 \tau_\nu. \tag{4.6}$$

That is, an extinction of 1 magnitude is equivalent to an optical depth of just under 1. Along the Galactic plane, the typical extinction at V band is about 1 magnitude per kiloparsec and thus the Galactic Center, at 8.2 kpc, is heavily obscured at optical wavelengths, $\tau_V \gg 1$.

A useful rule of thumb, apparent from Figure 4.2, is that the extinction in the infrared K band (2.2 μm) is approximately one-tenth that in the optical V band,

$$A_K \approx 0.1 \, A_V. \tag{4.7}$$

Thus we can see through to the Galactic Center in the near-infrared, $\tau_K \simeq 1$ (see Figure 1.1), although there are still patchy regions of obscuration.

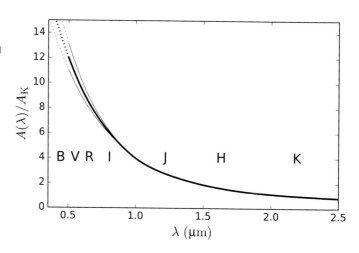

Fig. 4.2. The extinction curve for $R_V = 3.3$ (heavy line) and 3.6, 3.0 (upper and lower light lines respectively). The letters represent the central wavelengths of optical (BVRI) and near-infrared (JHK) filters commonly used in photometric observations.

The extinction curve in Figure 4.2 monotonically decreases with an approximate power law from optical to near-infrared bands,

$$A(\lambda) \propto \lambda^{-1.7}. \tag{4.8}$$

There are features on top of this smooth dependence due to resonances with compounds such as graphite in the ultraviolet (0.2175 μm) and silicate minerals in the mid-infrared (most prominently at 9.7 μm). Molecular bonds and, in cold regions, ice bands from water, carbon dioxide, and other molecules produce additional features throughout the near- to far-infrared. We discuss these in more detail in Chapter 7 but focus for now on the implications of the broad wavelength dependence.

4.2 Mie Theory

Our everyday experience tells us that an object casts a shadow over a region equal to its projected area. For example, a sphere with radius a would block light over an area πa^2. However, there are also diffraction effects around the edges and we will see that these are equally important. Furthermore, for small objects, there are more subtle effects due to the transmission of light through the objects that we must take into account.

Interstellar dust strongly blocks optical light but is more transparent in the near-infrared. If dust grains were macroscopic, they would cast the same shadow in the near-infrared as in the optical. The observed wavelength dependence of extinction therefore implies that grain sizes must be similar to this wavelength range. To be more quantitative, we define the **extinction efficiency**,

$$Q_{ext}(\lambda) = \frac{\text{extinction cross-section}}{\text{geometric cross-section}}. \tag{4.9}$$

We further divide this into scattering and absorption components,

$$Q_{ext} = Q_{sca} + Q_{abs}. \tag{4.10}$$

Mie (pronounced "me") first calculated the functional forms, $Q_{sca}(\lambda)$ and $Q_{abs}(\lambda)$, for a uniform sphere over a century ago. The solution depends on the dimensionless size, a/λ, and the index of refraction, $m(\lambda)$, of the sphere. m is a complex number where the real part slows the phase speed of the wave relative to vacuum as it passes through the material. This relates to refraction or scattering of the wavefront, and affects Q_{sca}. The imaginary part corresponds to absorption of the wave and mainly affects Q_{abs}.

Figure 4.3 shows the solution for a constant $m = 1.3 - 0.05i$. The simplest curve to understand is Q_{abs} (dotted line), which tells us that the absorption cross-section is low for particles that are smaller

Fig. 4.3. Extinction, scattering, and absorption efficiencies for a sphere with radius a and complex refractive index $m = 1.3 - 0.05i$ at wavelength λ.

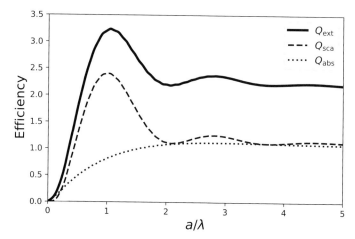

than the observing wavelength, and that it increases to the geometric cross-section for large particles. The latter is our everyday experience alluded to at the start of this section.

The scattering curve, Q_{sca} (dashed line), is also very small for small particles, and indeed is smaller than Q_{abs} for $a/\lambda \lesssim 0.2$ here, but increases much more steeply with size such that the scattering cross-section is greater than the geometric cross-section for particles that are comparable in radius to the observing wavelength. Q_{sca} then decreases and, with some smaller wiggles, asymptotes to about unity for larger particles. This is due to diffraction around the object with the result that the total (scattering plus absorption) extinction efficiency for a large object is *twice* its geometric size. For macroscopic objects, this apparent contradiction with everyday experience is known as the extinction paradox, with the reconciliation that the diffraction around the edges is so slight that this "scattered" component enters our eyes.

Going back to objects with sizes comparable to the wavelength, we see a sharp peak in Q_{sca} that is due to strong coupling between the electromagnetic wave and the particle. Heuristically, we can consider the interaction of a photon passing through the particle with others that pass nearby (Figure 4.4). The first photon is delayed due to its passage through the particle. This curves an incident plane wavefront and therefore scatters the light.

The full calculation requires solving Maxwell's equations, but this simple picture shows that the efficiency depends not only on a/λ but also on m. You are encouraged to explore this in the question section at the end of this chapter and empirically verify that the scattering efficiency peaks at

$$|m - 1|\frac{a}{\lambda} \simeq 0.3. \qquad (4.11)$$

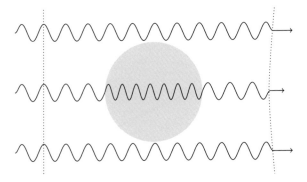

Fig. 4.4. A photon passing through a particle is delayed relative to photons on parallel, non-intersecting paths. This produces curvature in the wavefront, shown by the dotted lines.

Rising up to this peak value from long wavelengths, $Q_{sca} \propto 1/\lambda^4$. This dependence is known as Rayleigh scattering and explains why both the sky and reflection nebulae, such as the Pleiades, are blue. Another example is the Orion nebula, which is visible as a fuzzy patch to the naked eye or binoculars, but which appears predominantly stellar in near-infrared images (Figure 4.5). Dust in the diffuse ISM has $|m - 1| \simeq 0.3$ so the decrease in nebulosity at J band tells us that most of the grains have sub-micron sizes.

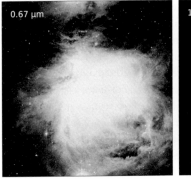

0.67 μm

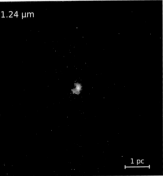

1.24 μm

1 pc

Fig. 4.5. The Orion nebula as seen in the optical (left) and near-infrared (right). Even though the wavelength is only two times longer on the right hand side, there is a dramatic difference in nebulosity due to the λ^{-4} scattering dependence. The images are taken from the Digitized Sky Survey (red plate) and Two Micron All Sky Survey J-band respectively and are one degree across on each side.

We have only discussed the wavelength dependence but there is also an angular dependence in the scattering, called the phase function. This is a more advanced topic and we note only that ISM grains scatter light more efficiently in the forward direction, toward the observer.

4.3 Grain Size Distribution

In practice, ISM dust comes in a range of sizes. The decline of extinction with wavelength shows that there are more smaller grains than larger ones. We can use Mie theory to fit the extinction law and thereby constrain the size distribution.

Fig. 4.6. Extinction cross-section versus wavelength (log scale) for different grain sizes, shown in the legend. The thick solid line shows the sum of the contributions from each of the three grain sizes and represents the total extinction that the mixture of dust particles provides.

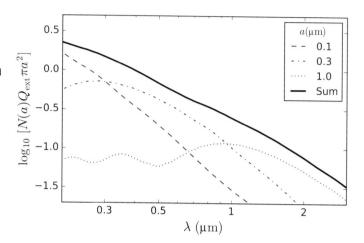

Figure 4.6 shows the extinction cross-section as a function of wavelength for three different particle sizes. Each curve is effectively the same as in Figure 4.3 but with the x-axis reversed and on a logarithmic scale. The smaller particles dominate at short wavelengths and the larger particles at longer wavelengths. The sum, multiplied by the number of particles at each size, is shown as the solid bold line and is the total extinction produced by the particle ensemble.

Varying the particle size distribution, $N(a)$, changes the slope of the total extinction. Only three sizes are shown here but the methodology can readily extend to many more and model a continuous size distribution such as a power law. For the observed ISM extinction law, the inferred numbers of dust grains with different sizes vary as $N \propto a^{-3.5}$. This has the interesting property that most of the surface area is in small grains but most of the mass (for uniform density) is in large grains. This is the reason that small amounts of dust have a large effect in the ISM, but it presents difficulties when trying to measure the dust mass.

4.4 Column Density

Due to our perspective that projects objects onto the plane of the sky, a fundamental quantity in ISM studies is the **column density**, defined as the number of particles per unit area along the line of sight. We describe how to measure this for dust here and for different components of the gas in subsequent chapters.

4.4.1 Absorption

The similarity of the extinction law toward different lines of sight, at least beyond $\sim 0.7\,\mu m$, tells us that the refractive index and size distribution of grains are quite uniform in most of the ISM. The amount of extinction and reddening therefore depends mainly on the amount of dust along the line of sight.

Dark clouds are extended objects and we can statistically estimate the amount of extinction by comparing the number of stars toward the cloud with the number away from it. This was first carried out in the 1940s using photographic plates of small, isolated clouds that are eponymously known as Bok globules. This star-counting technique naturally works best on nearby objects with little foreground star contamination. Optical observations are also naturally restricted to relatively diffuse objects that do not completely block all background stars.

In the 1990s, as infrared detector arrays became large enough to map large areas of the sky, the star-counting technique was extended to longer wavelengths to study more (visually) opaque regions. A near-infrared image of a dark cloud, such as one of the panels in the bottom row in Figure 4.1, is divided up into a set of regions where the number of stars in the background, off the cloud, is high and generally where there are at least a few stars detected in regions toward the cloud. Let the number of stars in the background be N_0, in an angular area Ω_0, and the number of stars in a region on the cloud be N_1 in Ω_1. The sensitivity of the observations determines the luminosity, L_0, of the faintest objects that are detected in the unextincted background. The intrinsic luminosity, L_1, of the faintest detected sources through the cloud will be higher, however, due to the extinction by the cloud. With knowledge of the luminosity distribution of the stars, from off-cloud observations or using Galactic models, we can then compare the observed numbers to derive the opacity. If the cumulative distribution of stellar number per unit area follows a power law, $N(> L) \propto L^{-\gamma}$, then

$$\frac{N_0/\Omega_0}{N_1/\Omega_1} = \left(\frac{L_0}{L_1}\right)^{-\gamma}. \tag{4.12}$$

If the cloud has optical depth τ, then $L_1 = L_0 e^{\tau}$, and the extinction at the observing wavelength, λ, is then readily derived,

$$A_\lambda = \frac{1.086}{\gamma} \ln\left(\frac{N_0 \, \Omega_1}{N_1 \, \Omega_0}\right). \tag{4.13}$$

The error in the measurement is largely Poisson statistics and the resolution of the extinction map is set by the area over which stars are counted. There is a balance, then, between obtaining high numbers, N_1, to reduce counting uncertainties and measuring the structure of the cloud over small scales, Ω_1.

The extinction can be measured more precisely toward individual stars through their reddening and use of the extinction curve. This requires knowing intrinsic colors but, for all but the coolest sources, these do not vary greatly beyond 1 μm because the (approximately blackbody) emission is in the Rayleigh–Jeans limit, $L_\nu = 4\pi R_*^2 B_\nu(T_*) \approx 8\pi R_*^2 kT_* \nu^2/c^2$, with the same frequency dependence for all stars. In the magnitude system, the color would therefore be identical for all stars, and equal to zero as defined for an A0 star. An additional advantage of the near-infrared is the reduced sensitivity to the value of R_V in the conversion from color to extinction (Figure 4.2). Deep images of a dark cloud at H and K bands can therefore provide measurements of the extinction along numerous lines of sight. The error in the measurement is largely in the range of intrinsic colors and the resolution depends on the density of lines of sight through the cloud.

These measures of extinction can be related to the amount of dust grains using Mie theory as described above. There are empirical conversions between extinction and gas surface density in the ISM that we will explore in Chapter 5.

4.4.2 Emission

Dusty clouds strongly absorb ultraviolet and optical photons. Their energy is conserved and released at longer wavelengths as the absorbing dust grains are cooler than the source of emission. Consequently, optically dark cores such as Barnard 68 emit in the far-infrared (Figure 4.7).

The strength of the emission depends on both the amount and temperature of the dust. Assuming for now a uniform dust temperature, T_d, we can use Equation 3.12 with no background term to write down the specific intensity,

$$I_\nu = B_\nu(T_d)\left(1 - e^{-\tau_\nu}\right). \tag{4.14}$$

This modified blackbody is often referred to as a **graybody**.

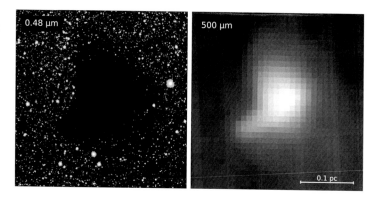

Fig. 4.7. The Barnard 68 core as seen in absorption in the optical (left) and emission in the far-infrared (right). Each map extends over 7.5×7.5. The images are taken from the Pan STARRS and Herschel SPIRE archive respectively.

If we consider the passage of light through a cloud with a number density of dust grains, n_d, each particle presents an extinction cross-section, $Q_{ext}(v)\pi a^2$, and the optical depth along a pathlength ds is

$$d\tau_v = \kappa_v ds = n_d Q_{ext}(v)\pi a^2 ds. \qquad (4.15)$$

The convention for dust (but not for atoms or molecules), is to factor out the mass density, $\rho_d = n_d m_d$, in the absorption coefficient,

$$\kappa_v = \rho_d \kappa_v^{dust}, \qquad (4.16)$$

where m_d is the mass of an individual dust grain and $\kappa_v^{dust} = Q_{ext}\pi a^2/m_d$ is the cross-section per unit mass with units $m^2\,kg^{-1}$.

Far-infrared wavelengths are much larger than individual dust grains. Thus the interaction of the radiation and material is weak and the optical depth is low. From Equation 4.14, we can therefore approximate $I_v \simeq B_v(T_d)\tau_v$, which implies that the observable flux is

$$F_v = \int_\Omega I_v\, d\Omega \simeq \int_V B_v \rho_d \kappa_v^{dust} dV/d^2 = \kappa_v^{dust} B_v M_d/d^2. \qquad (4.17)$$

The steps follow by converting the integral over solid angle to one over area, where d is the distance to the source, and then to combining the integral over pathlength to convert from area to volume. The final step assumes that κ_v^{dust} and T_d are uniform over the observed region and motivates the factoring out of the density of grains from their optical properties.

As the emission is optically thin, the observed flux measures the total grain effective surface area, which we then convert to mass via κ_v^{dust}. This can be determined via the relation between κ_v^{dust} and Q_{ext} above. Calculations for different grain size populations and compositions show an approximate power law form,

$$\kappa_v^{dust} = \kappa_0^{dust}\left(\frac{v}{v_0}\right)^\beta, \qquad (4.18)$$

for $v < 1200\,GHz$ ($\lambda > 250\,\mu m$). At long wavelengths, the Planck function is in the Rayleigh–Jeans limit, $B_v(T_d) \approx 2kT_d v^2/c^2$. Thus the flux has an approximate power law dependence, $F_v \propto v^{2+\beta}$. Galactic multi-wavelength surveys show a spectral dependence, $\beta \approx 1.7$. Coupled with other mass measurements, which we will discuss later, the normalization at $\lambda = 1\,mm$ is $\kappa_{1\,mm}^{dust} \approx 0.03\,m^2\,kg^{-1}$. Combining the uncertainties in the mass absorption coefficient, temperature, and distance, thermal dust emission measurements provide reasonably accurate mass estimates of objects both within our Galaxy and in others to within a factor of a few.

The thermal emission from ISM dust is weaker than from a blackbody of the same temperature because the grains are small and emit inefficiently at long wavelengths. This is the flip side of their low extinction due to the weak interaction with long-wavelength radiation. However, protoplanetary disks are found to have a significantly lower $\beta \approx 1$. This can be explained by grain growth from (sub-)micron sizes in cloud cores such as Barnard 68 to millimeter sizes and beyond in disks. The dense, quiescent regions in disks allow dust particles to grow by a factor of over 1000 in size through agglomeration and buildup of icy mantles from small grains to "pebbles" that are the building blocks of planets. Since most of the mass is in the large particles, the corresponding conversion from millimeter emission to dust mass in disks is about an order of magnitude larger than in the diffuse ISM, $\kappa_{1\,\mathrm{mm}}^{\mathrm{dust}} \approx 0.3\,\mathrm{m}^2\,\mathrm{kg}^{-1}$. The implications of this for planet formation are explored further in Chapter 9.

4.5 Temperature

The emission from a dust cloud depends not only on its mass but also its temperature. If we consider a spherical particle with radius a at a distance d from a star with luminosity L_*, its equilibrium temperature can be calculated by balancing the energy absorbed by the particle per unit time,

$$\dot{E}_{\mathrm{abs}} = \frac{L_*}{4\pi d^2}\,\pi a^2, \qquad (4.19)$$

with the energy radiated per unit time,

$$\dot{E}_{\mathrm{em}} = 4\pi a^2 \sigma T_{\mathrm{bb}}^4, \qquad (4.20)$$

where σ is the Stefan–Boltzmann constant. Under these assumptions of perfect absorption and emission (i.e., a blackbody), the particle temperature is

$$T_{\mathrm{bb}} = \left(\frac{L_*}{16\pi \sigma d^2}\right)^{1/4}. \qquad (4.21)$$

The typical distance between stars in the solar neighborhood is about 2 pc so the radiation field is very weak and temperatures are very low. The average flux of starlight is $G_0 = 1.6 \times 10^{-6}\,\mathrm{W\,m}^{-2}$ in the far-ultraviolet range 6–13.6 eV where most heating occurs, a quantity that is called the Habing field. Bathed in this isotropic field, the equilibrium temperature of a blackbody would be $(G_0/\sigma)^{1/4} = 2.3\,\mathrm{K}$, which is a little lower than that of the cosmic background radiation, $T_{\mathrm{CBR}} = 2.7\,\mathrm{K}$.

However, as we have shown in this chapter, interstellar dust grains are not perfect blackbodies. If we take into account their absorption efficiency, the above equations generalize to

$$\dot{E}_{\text{abs}}(a) = \int \frac{L_\nu}{4\pi d^2} Q_{\text{abs}}(a, \nu)\pi a^2 \, d\nu, \tag{4.22}$$

$$\dot{E}_{\text{em}}(a) = \int \pi B_\nu(T_{\text{gr}}) Q_{\text{abs}}(a, \nu)4\pi a^2 \, d\nu. \tag{4.23}$$

The πB_ν term is the emergent flux from each point on the grain surface and comes from integrating the blackbody intensity in projected angular area over a half-sphere,

$$F_\nu = \int B_\nu d\Omega = B_\nu \int_0^{2\pi} d\theta \int_0^{\pi/2} \cos\phi \sin\phi \, d\phi = \pi B_\nu. \tag{4.24}$$

We then multiply this term by the effective surface area and integrate over frequency to obtain the total energy radiated from a grain.

The same efficiency, Q_{abs}, is in each term because, at any given frequency, a grain will emit with the same efficiency that it absorbs light. However, the weighting in these integrals is very different. The stellar radiation term peaks in the optical where dust has a high absorption efficiency, $Q_{\text{abs}} \simeq 1$, but the dust is much cooler due to geometric dilution of the radiation field and the blackbody term peaks in the infrared where the absorption (and therefore emission) efficiency is low, $Q_{\text{abs}} \ll 1$. With these considerations, we must therefore solve

$$\int \pi B_\nu(T_{\text{gr}}) Q_{\text{abs}}(a, \nu) \, d\nu = \frac{L_*}{16\pi d^2}. \tag{4.25}$$

This can be iteratively solved to determine the grain temperature at different sizes, shown in Figure 4.8. Small grains are efficient absorbers of starlight but inefficient radiators at long wavelengths. They are therefore significantly warmer than the blackbody calculation above. For ultra-small grains on the left of the plot, the absorbed energy rate is low but the emitted photon energy is high and a grain emits stochastically through molecular bond vibrational modes that produce infrared

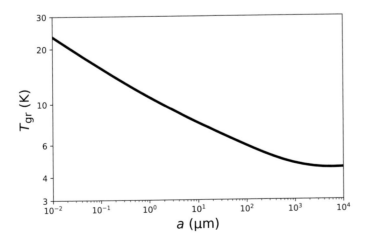

Fig. 4.8. Equilibrium temperature for different size dust grains heated by the mean interstellar radiation field.

lines corresponding to much higher temperatures than this equilibrium calculation indicates. At the other extreme, on the right side of the plot, centimeter-sized particles are large enough to behave as blackbodies and temperatures plateau. Such larges sizes are much greater than found in the ISM but grains do grow to this level in protoplanetary disks.

Fits to far-infrared maps of the Galaxy show that the dust SED peaks at a few hundred microns and has an average temperature of 19 K, corresponding to a typical size $a \sim 0.1$ μm. Temperatures are naturally higher in stronger radiation fields such as the vicinity of luminous stars or the dusty environments of protostars (Chapter 9).

4.6 Polarization

Maxwell's equations tell us that a light wave consists of electric and magnetic fields oscillating at right angles to each other and the direction of propagation. If the orientation of the electric fields is uniformly distributed the light is unpolarized, but if there is a preferred direction for the orientation the light is polarized. The degree of polarization quantifies this preference and ranges from 0% for unpolarized light to 100% when the electric fields are fully aligned in a single direction. The polarization angle specifies the preferred orientation.

The decomposition of light into its polarized components is described by the **Stokes parameters**. These can be measured through a variety of techniques, such as Wollaston prisms that split optical light into two orthogonally polarized beams. In radio telescopes, the detector is fundamentally a line conductor that measures the alternating current induced by the radiation. This is inherently one dimensional and therefore sensitive to only one direction of polarization. This has the disadvantage that two orthogonal detectors (and receivers) are required to measure the total intensity, but the advantage that the polarization is readily determined by comparing the strength of the signal between the two.

Polarization encodes additional information to the wavelength and intensity of the light. In the context of dust, scattering and absorption may induce polarization and emission can be intrinsically polarized. To conceptually understand this, first note that we can decompose unpolarized light into the sum of two orthogonal linearly polarized waves with equal intensity, and second that light interacts with matter predominantly through the electric field.

The effect of scattering is illustrated in Figure 4.9. The electric field in the outgoing ray must be perpendicular to the new direction of propagation, so the component lying in a plane defined by the incoming and outgoing lights rays is reduced. This is most readily visualized in

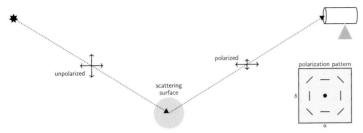

Fig. 4.9. Scattering-induced polarization by a dust grain. The arrows on each light ray indicate the electric field direction and are of equal size for the incoming unpolarized light, but the component in the plane defined by the incoming and outgoing light rays is attenuated so the telescope sees light polarized at an angle perpendicular to the scattering surface.

the extreme case of scattering at right angles, where the electric field would be exactly in the outgoing direction and cannot propagate. As the electric field perpendicular to the plane is unaffected, the outgoing light ray is polarized.

The polarization angle is perpendicular to the plane and a telescope that observes along multiple lines of sight would map a circular polarization pattern centered on the source. The degree of polarization is typically a few percent per visual magnitude of extinction. In regions of very high extinction, however, there may be multiple scattering events which scramble the polarized signal. For the single scattering case, we can use its circular symmetry to locate highly obscured sources or, by observing at a particular polarization angle perpendicular to a given source, mitigate confusion from other sources (analogous to using polarized sunglasses on water to avoid glare from reflected sunlight and see below the surface).

The absorption and emission from dust grains is also found to be polarized in many cases. In this case, the effect is due to aligned, elongated dust grains which either block or produce light preferentially in one direction, as illustrated in Figure 4.10.

Using Mie theory, we modeled the extinction efficiency with a single parameter, a, for the radius of a spherical dust grain. If the grains are non-spherical, however, the cross-section for absorption and emission is greater for light with its electric field orientated in the elongated direction. If the dust grains are randomly orientated, there is no net polarization signal, but if they are somehow aligned, the absorbed and emitted light will be polarized in orthogonal directions.

Though not magnetic themselves, silicate dust grains are paramagnetic and are attracted to an external magnetic field through an induced interior field. The grains therefore follow the ISM magnetic field lines, **B**, and become aligned as they lose energy through radiation

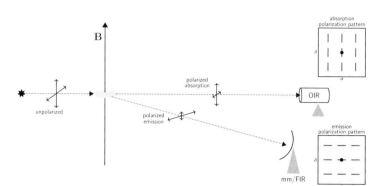

Fig. 4.10. Polarized absorption and emission due to elongated dust grains aligned by a magnetic field, **B**. An optical or infrared (OIR) telescope measures absorbed light from an unpolarized source of emission. The dust grain cross-section is larger for the horizontal electric wave so the telescope measures a polarization pattern that is parallel to the **B** field. The emission, however, is polarized along the direction of grain elongation, so a telescope that measures the thermal dust emission at millimeter or far-infrared (mm/FIR) wavelengths will measure a polarization pattern that is perpendicular to the **B** field.

but conserve angular momentum, spinning around the smallest axis of their moment of inertia. This arrangement leads to preferential absorption along the longer axis and an observed polarization pattern in the outgoing light that is parallel to **B**. The emission is polarized in the elongated direction of the grains and perpendicular to **B**.

Thus dust absorption observations at short wavelengths, or emission at longer wavelengths, reveal the presence and geometry of magnetic fields in the ISM. The degree of polarization is generally small, at most 10%, and sensitive to the grain shape and composition. Because of the dependence on grain optical properties, the strength of the magnetic field cannot be determined from the degree of polarization, but it is related to the variation in polarization angle from point to point. The degree to which grains are aligned is a balance between field strength and collisions with gas particles. The **Chandrasekhar–Fermi method** relates the magnetic field with the ratio of the gas linewidth to the dispersion in polarization angle. We will discuss the role of magnetic fields in the dynamics of molecular clouds and star-forming regions later in the book.

4.7 Cosmic Dust in the Laboratory

It is estimated that about 100 tonnes (10^5 kg) of cosmic dust hit the Earth each day. Millimeter and larger sized particles burn up in the atmosphere to produce flashes of light as meteors, and the largest may reach the ground as meteorites. Very small particles, just a few microns

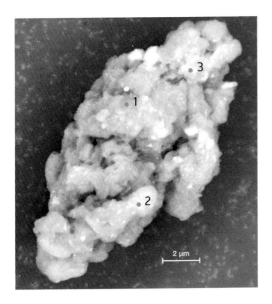

Fig. 4.11. Scanning electron microscope image of a cosmic dust particle captured by a high-altitude airplane. The numbers indicate three spots where the elemental composition was measured. From the NASA Astromaterials Acquisition and Curation Office curator.jsc.nasa.gov.

in size, survive entry and eventually drift down to the ground. High-altitude aircraft, flying in the stratosphere at ~20 km, can catch such particles on their way down and bring them in for detailed examination. Most particles have meteoritic compositions and are debris from rocky body collisions in the asteroid belt or relics of planet formation. A small fraction have anomalous isotopic signatures that indicate an interstellar origin. These provide a remarkable opportunity to study extraterrestrial material in the laboratory.

These captured cosmic dust particles have typical sizes of a few microns and can be imaged with scanning electron microscopes (Figure 4.11). This is on the larger end of the size distribution inferred from ISM observations and represents the current limit on reliable identification and characterization. The images typically show that the particles are amalgams of sub-micron grains that have come together in a random-walk process and have a high degree of porosity. Consequently, their effective surface area is much larger than a smooth object with the same size, and they play an essential role in providing a site for atoms to react and form molecules (Chapter 7).

The composition of such particles (and meteorites) can be studied in exquisite detail through the fascinating field of **cosmochemistry**. This reveals the process of ISM dust formation in the cool, expanding envelopes of evolved stars and in supernova ejecta (Figure 8.5). For inter-planetary dust grains, such studies provide unique insights into the origin of the Solar System.

Notes

The use of selective extinction, R_V, to parameterize the variations in the extinction law was introduced by Mathis et al. (1977). More recent work characterizing the extinction law from all-sky optical and infrared surveys can be found in Schlafly et al. (2016). The ISM grain size distribution, $N \propto a^{-3.5}$, is known as the MRN distribution after the authors of the paper, Mathis et al. (1977), based on their modeling of the extinction law. The Protostar and Planets series contains a review by Lada et al. (2007) of the infrared star count technique to measure the structure of dark clouds. Ossenkopf and Henning (1994) provide a commonly used tabulation of the mass absorption coefficient, κ_ν^{dust}, for different dust grain models and the expression in Equation 4.18 comes from Hildebrand (1983). The textbook, *Physics of the Interstellar and Intergalactic Medium*, by Draine (2011) provides additional detail about dust, and indeed all the other subjects in this book, at a more advanced level.

Questions

1. Look up the absolute magnitude of the Sun at V band. What would the apparent magnitude be for a solar twin at the Galactic Center? What would it be with dust assuming that the extinction along the Galactic plane is $1\,\mathrm{mag\,kpc^{-1}}$? Carry out the same calculations in the near-infrared at K band.

2. Estimate the angular area of the Barnard 68 core in Figure 4.7. Assume that the dust has a uniform temperature of $20\,\mathrm{K}$ and that the total dust mass is $10^{-2}\,M_\odot$. Plot the SED of the dust emission assuming a dust absorption coefficient described by Equation 4.18 and associated text. Overplot a blackbody with the same solid angle and temperature and explain the difference.

3a. Recreate Figure 4.3 using Mie theory. There are several publicly available codes available. I used and recommend github.com/scottprahl/miepython. Once you have successfully reproduced the plot, try varying the index of refraction and show that the scattering efficiency peaks at $|m - 1|a/\lambda \simeq 0.3$.

3b. Zoom into the left side of the plot and show that $Q_{abs} > Q_{sca}$ for small a/λ.

3c. Consider a population of dust grains with different sizes. Plot the extinction cross-section as a function of λ as in Figure 4.6 and explore how the resulting overall extinction curve depends on the grain size distribution.

Chapter 5
Atomic Regions

By mass, most of the Galactic ISM is atomic hydrogen and can be readily observed with radio telescopes. Far-infrared emission lines of oxygen and carbon take energy away from the clouds and regulate their temperature. Trace elements produce absorption features in the ultraviolet spectra of background stars and quasars. Their inferred abundances tell us not only about the composition of the gas but also of the accompanying dust.

In a recurring pattern of astronomers holding on to an old but confusing nomenclature, we use Roman numerals I, II, III, etc. to label ionization states, with I representing neutral, II representing a singly ionized state, and so on. Thus atomic regions are interchangeably called HI clouds. We discuss these here and ionized, or HII, regions in the next chapter.

5.1 The 21 cm Transition

The electron orbital energy level of the first excited state of hydrogen is 10.2 eV above the ground level. This is much greater than the average kinetic energy in the ISM and collisional excitation is therefore rare. Consequently HI lies in the ground state and does not interact with optical or infrared radiation.

Fortunately, HI can still be detected through the hyperfine splitting of the ground electronic state (Figure 5.1). The proton and electron each have intrinsic angular momentum, or spin, and their orientations can be aligned (parallel) or opposite (anti-parallel and the lowest energy state). The energy difference between the two is a minuscule 6 μeV, which corresponds to 1.42 GHz in frequency or 21 cm in wavelength.

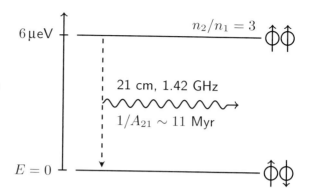

Fig. 5.1. Schematic of the hyperfine transition of the ground electronic state of hydrogen, due to the spin flip from parallel to anti-parallel alignment of the proton–electron pair.

The discovery of the HI 21 cm line is an archetypal example of the scientific method because its theoretical prediction by van der Hulst in 1945 led to the experimental design of an antenna and subsequent detection by Ewen and Purcell in 1951. This is unusual in astronomy since, as an observationally driven science, most discoveries tend to be theoretically explained after the fact. Regardless, its utility was quickly recognized for studying the ISM and Galactic structure. Today it is an important tool for studying nearby galaxies and it will be the way in which we see the first atoms in the Universe.

Because the transition is so low in energy, the spontaneous emission rate rate is very small,

$$A_{21} = 2.9 \times 10^{-15}\,\mathrm{s}^{-1}. \tag{5.1}$$

That is, an atom will typically remain in the excited state for about 11 Myr before decaying. This is much greater than the collisional time between atoms in an HI cloud. Thus the two hyperfine states are in collisional equilibrium and the level populations are in a Boltzmann distribution with excitation temperature equal to the kinetic temperature of the gas, $T_{\mathrm{ex}} = T$,

$$\frac{n_2}{n_1} = \frac{g_2}{g_1} e^{-E_{21}/kT} \simeq 3, \tag{5.2}$$

where the numerical value comes from the quantum degeneracies, $g_1 = 1, g_2 = 3$, and the approximation is because $E_{21}/k = 0.07\,\mathrm{K}$ is much less than ISM temperatures (and indeed that of the cosmic background radiation) so the exponential term is near unity.

5.2 Column Density

Although HI is in almost perfect collisional equilibrium, a small fraction of excited atoms decay through spontaneous emission. This allows us to

determine the number of hydrogen atoms atoms along the line of sight
from the strength of the 21 cm line. We begin with the general radiative
transfer solution (Equation 3.12) assuming a uniform temperature,

$$I_\nu = B_\nu(T(1 - e^{-\tau_\nu}) \simeq \frac{2kT\nu^2}{c^2}(1 - e^{-\tau_\nu}), \qquad (5.3)$$

where we have used the Rayleigh–Jeans approximation for the Planck
function since $h\nu \ll kT$.

The optical depth is the integral of the absorption coefficient along
the line of sight,

$$\tau_\nu = \int \kappa_\nu ds, \qquad (5.4)$$

where κ_ν is given by Equation 3.18 and, using the relations between the
Einstein coefficients, this becomes

$$\tau_\nu = \frac{3A_{21}hc^2}{32\pi \nu k} \frac{N_H}{T} \phi(\nu), \qquad (5.5)$$

where we introduce the hydrogen column density,

$$N_H = \int n_H ds, \qquad (5.6)$$

with units of m^{-2}. This is effectively a surface density expressed in
numbers of particles, rather than mass, per area. Note that $n_H = n_1 + n_2 = 4n_1$ because the system is in Boltzmann equilibrium.

The fundamental and atomic constants in Equation 5.5 give the scal-
ing factor $3A_{21}hc^2/32\pi \nu k = 2.6\times10^{-19}$ K m^2 s^{-1}. The peak of the line
profile $\phi(\nu) \sim 1/\Delta\nu$ where $\Delta\nu$ is the line width. The typical motions
in atomic clouds are several km s^{-1} corresponding to $\Delta\nu = \nu\Delta v/c$
$\sim 10^4$ Hz. Unless the combination $N_H/T \gtrsim 4 \times 10^{22}$ m^{-2} K^{-1}, the
emission is therefore optically thin and the specific intensity is

$$I_\nu = \frac{3A_{21}}{16\pi} N_H h\nu \, \phi(\nu). \qquad (5.7)$$

This has the basic form of a rate times a number per area times an
energy with a line profile describing how the intensity is distributed in
frequency. There is no dependence on temperature or collisional rate
because the levels are effectively fixed in Boltzmann equilibrium. Given
the low Einstein A value, the fact that the 21 cm line is detectable at all
is only due to the enormous number of neutral hydrogen atoms.

Since $B_\nu(T) = 2k\nu^2 T/c^2$ in the Rayleigh–Jeans limit, centimeter
wavelength radio astronomy generally expresses the specific intensity
as a **brightness temperature**,

$$T_B \equiv \frac{c^2}{2k\nu^2} I_\nu. \qquad (5.8)$$

For optically thick thermal emission, $I_\nu = B_\nu(T_{ex})$, so T_B is a direct measurement of the excitation temperature of a source. Indeed, a radio receiver is calibrated by placing hot and cold loads of known temperatures in the optical path and scaling the output to those values. Spectra are generally plotted as brightness temperature versus (Doppler-shifted) velocity. For the optically thin case here, $I_\nu \neq B_\nu$ and T_B relates to the column density. Integrating over the line profile leads to the following relation,

$$N_H = 1.82 \times 10^{26} \, \text{m}^{-2} \left[\frac{\int T_B dv}{\text{K km s}^{-1}} \right]. \qquad (5.9)$$

The term in square brackets, with the uniquely radio units of K km s^{-1}, is the **integrated line intensity** and is the simple observable of the area under an HI spectrum.

The column density is independent of distance, but converting to mass requires integrating over the physical area that relates to the solid angle and distance, d, to the cloud, $M_H = N_H m_H \Omega d^2$. In addition, hydrogen only makes up 74% of the overall mass (about one in eleven atoms is helium) so the total mass should be augmented by a factor $\mu = 1/0.74 = 1.35$.

A map of the HI column density is included in the Galaxy multi-wavelength montage in Figure 1.1. There is strong emission along the Galactic plane and wispy features above and below. The large shell-like structures are the edges of kiloparsec-sized bubbles created by clustered supernovae. We will discuss the dynamics of these objects in Chapter 8. There is tremendous granularity in these maps, especially when we note that each spatial pixel is actually a spectrum and contains velocity information. A map can therefore be made at any particular velocity, or integrated over a range of velocities, and the kinematics of the Galaxy can be revealed. A smaller part of the Galactic plane at a particular velocity is shown in Figure 5.2.

By comparing the 21 cm emission at different velocities, it was possible to associate atomic clouds with star-forming regions. Distances could then be calculated from the stellar parameters, and the variation of Galactic rotation with radius, $v_{rot}(R)$, was determined in the 1970s. The rotation curve is flat, $v_{rot} \sim 220 \, \text{km s}^{-1}$ out to the edge of detectable emission, indicating gravitational effects from massive amounts of unseen material, i.e., the dark matter halo (see Chapter 10). These were hard won advances 50 years ago, but the pace of technology now makes the construction of a basic HI telescope and mapping Galactic rotation a realistic undergraduate project!

Working in the opposite direction, once the Galactic rotation curve was established, the radial velocity of any 21 cm feature could be converted to an approximate distance and HI column densities transformed

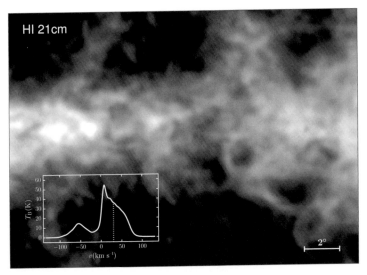

HI 21cm

Fig. **5.2.** Galactic 21 cm HI line emission from the HI4PI survey, covering longitude $40° < l < 60°$, latitude $-5° < b < 5°$, at a velocity of 30 km s^{-1}. The inset shows the mean spectrum of this region in units of brightness temperature versus velocity. The vertical dotted line shows the velocity slice shown in the image.

into masses. This led to an estimate of the total atomic mass (hydrogen plus helium) in the Galaxy of $\sim 10^{10}\,M_\odot$, which is about an order of magnitude lower than the total mass in stars. The average volume density can also be inferred from the observed column density once the size scales are known. It is hard to define the boundaries of intrinsically nebulous ISM features but the typical densities for atomic clouds range over $n_H \sim 10^{6-8}\,\text{m}^{-3}$.

5.3 Temperature

The simplicity of the 21 cm column density calculation is due to the HI hyperfine states always being in Boltzmann equilibrium. Consequently, the emission depends only on the amount of neutral hydrogen along the line of sight and not on how dense or warm it is. So how then can we measure the temperature of atomic gas?

The answer lies in the inverse temperature dependence of the opacity (Equation 5.5). Cold gas absorbs more strongly than warm gas, ultimately due to the stimulated emission term B_{21} in Equation 3.17. We therefore need to consider line absorption. The simplest geometry is shown in Figure 5.3 where two neighboring lines of sight pass through an HI cloud with a uniform kinetic temperature T_H and optical depth τ. The first line of sight measures cloud emission only. Dropping the ν dependence for clarity and using the definition of brightness temperature in the Rayleigh–Jeans formulation for the radiative transfer (Equations 5.3, 5.8), we have

$$T_{B,1} = T_H (1 - e^{-\tau}). \tag{5.10}$$

Fig. 5.3. Schematic of an absorption experiment to measure the temperature of an atomic cloud. The first line of sight measures the emission from the cloud only, illustrated by the spectrum in the top left. The second line of sight contains a background source with continuum temperature T_c. The difference between the two lines of sight reveals a dip in the continuum emission due to absorption by the cloud, illustrated by the spectrum in the bottom right.

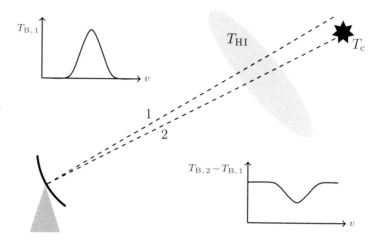

Fig. 5.3. Schematic of an absorption experiment to measure the temperature of an atomic cloud. The first line of sight measures the emission from the cloud only, illustrated by the spectrum in the top left. The second line of sight contains a background source with continuum temperature T_c. The difference between the two lines of sight reveals a dip in the continuum emission due to absorption by the cloud, illustrated by the spectrum in the bottom right.

However, the second line of sight includes an additional component, with brightness temperature T_c, from the background source,

$$T_{B,2} = T_H \left(1 - e^{-\tau}\right) + T_c e^{-\tau}. \tag{5.11}$$

The background source is typically a quasar that emits over a continuum so T_c can be measured at neighboring frequencies away from the HI line. Thus the two equations can be solved to determine the two unknowns,

$$\tau = \ln\left(\frac{T_c}{T_{B,2} - T_{B,1}}\right),$$
$$T_H = \frac{T_{B,1}}{1 - e^{-\tau}}. \tag{5.12}$$

Just as the emission is a spectral feature that encodes information about the HI velocities, so is the optical depth, as shown schematically by the insets in Figure 5.3. The difference between the two lines of sight (in practice this is most accurately measured using an interferometer) dips below the continuum level at the velocities where the HI absorbs.

Actual data are, of course, more complicated. An example is shown in Figure 5.4. The emission is broad and approximately Gaussian in shape but we see several distinct absorption features. The narrowness of the absorption lines compared to the broadness of the emission profile is due to the greater absorbing efficiency of cold gas. The multiple features show that there are several distinct components, with different temperatures, along the line of sight.

This non-uniformity complicates the interpretation of the temperature measurement described above. We can readily generalize

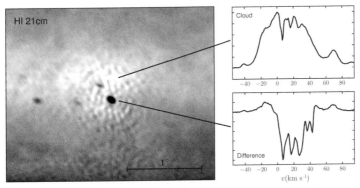

Fig. 5.4. A region of the Southern Galactic Plane Survey showing HI emission and absorption toward a compact radio source (an HII region) at the center. The top spectrum on the right hand side shows the cloud emission and the lower spectrum shows the difference when looking directly on source. The mottled appearance in the image is due to artifacts from the bright continuum source. A few other fainter absorption features are also visible in the image. Their elliptical shapes are due to the elongated interferometric beam.

Equation 5.10 to include two clouds a and b with kinetic temperatures T_a, T_b,

$$
\begin{aligned}
T_{B,1} &= T_a(1 - e^{-\tau_a}) + T_b(1 - e^{-\tau_b})e^{-\tau_a}, \\
T_{B,2} &= T_a(1 - e^{-\tau_a}) + T_b(1 - e^{-\tau_b})e^{-\tau_a} + T_c e^{-\tau_a}e^{-\tau_b}.
\end{aligned}
\tag{5.13}
$$

We have assumed here, without loss of generality, that a is in front of b. Now, if we cannot distinguish between the two clouds through their velocity profiles and try to infer the cloud properties as if they were a single entity using Equation 5.12, we would derive an optical depth that is the sum along the line of sight,

$$
\tau = \ln\left(\frac{T_c}{T_{B,2} - T_{B,1}}\right) = \tau_a + \tau_b.
\tag{5.14}
$$

However, the inferred HI temperature is a more complicated combination of the two clouds,

$$
T_H = \frac{T_{B,1}}{1 - e^{-\tau}} = \frac{T_{B,1}}{1 - e^{-[\tau_a + \tau_b]}}.
\tag{5.15}
$$

For small optical depths, $\tau_a, \tau_b \ll 1$, this simplifies to the opacity weighted mean,

$$
T_H \simeq \frac{T_a \tau_a + T_b \tau_b}{\tau_a + \tau_b}.
\tag{5.16}
$$

Note that the product of temperature times optical depth is proportional to the column density (integrate Equation 5.5 over the line profile) so this equation is equivalent to saying that the total column density of the two clouds is the sum of each one, $N_{tot} = N_a + N_b$. That intuitively makes sense given the low optical depths, but we can then restate Equation 5.14 to

$$\frac{N_{tot}}{T_H} = \frac{N_a}{T_a} + \frac{N_b}{T_b}. \qquad (5.17)$$

That is, the inferred temperature of the composite is the weighted harmonic mean of the individual clouds,

Early 21 cm maps of the Galaxy, in the 1960s, revealed widespread HI emission. Careful absorption measurements found temperatures that varied by almost two orders of magnitude, from $\sim$100 K to $\sim$8000 K. This is harder than it seems because cold gas is a much stronger absorber than warm gas. Nevertheless, once the presence of two components was established, new questions arose: why are some regions cold, others warm, and how are they distributed? This has an elegant answer but the details remain a pressing topic of current research.

5.4 The Two-Phase Neutral Medium

The temperature of the air in the room where you are reading this is determined via conduction or convection from a surface. The situation in the ISM is of course quite different and the temperature of an atomic cloud depends on the production of photoelectrons. Ionization of a species will produce a fast moving electron that can redistribute its kinetic energy with the rest of the gas and increase its temperature. This is countered by electron collisions that excite internal energy levels of other species which can then radiate away some of the kinetic energy and cool the gas. The discreteness of these various quantum processes naturally leads to distinct phases in the atomic ISM. There are many details in a full description so we heuristically outline the basic principles here.

The ionization potential of hydrogen is 13.6 eV, corresponding to far-ultraviolet radiation. As testified by the widespread 21 cm emission from neutral gas, this is is well above the typical energy of diffuse starlight in the ISM. However, optical light can produce photoelectrons from dust grains. The released electron has a kinetic energy equal to the photon energy minus the work function of the dust grain and is typically a few eV. As we will see later, the number of dust grains is proportional to the number of hydrogen atoms and the total heating of the cloud per unit volume can therefore be represented as

$$\text{Heating} = n_H \Gamma, \qquad (5.18)$$

with units of $W\,m^{-3}$, where Γ scales with the intensity of the radiation field.

Fig. 5.5. CII 158 μm line emission in the Galaxy. The map size is −180° to +180° in Galactic Longitude and −60° to +60° in Galactic Latitude (the same as the multi-wavelength panoramas in Figure 1.1). The brightness represents the line intensity and varies from 3 to 200 nW m⁻² sr⁻¹. The data are from all-sky maps created by the Cosmic Microwave Background Explorer.

The energy of a photoelectron is insufficient to excite the first electronic state of hydrogen at 10.2 eV (see Chapter 3). Thus a HI–e collision exchanges kinetic energy. Because of the large mass ratio, many interactions are required for equipartition and interactions with other species will also occur. The next abundant species are carbon, nitrogen, and oxygen. Carbon differs from the others in having an ionization potential, 11.3 eV, below that of hydrogen and can therefore be at least partially ionized in HI clouds. Electrostatic attraction implies that the cross-section for CII–e collisions is relatively large and, moreover, the CII ground state is split by $E_0 = 0.008$ eV (this is known as fine structure and is due to relativistic and other corrections to the Hamiltonian describing the electron orbitals). Because CII is relatively abundant and readily excited, it siphons off much of the photoelectron kinetic energy. Indeed, the similarity between the temperature $T_0 = E_0/k = 92$ K and the low end of observed HI temperatures demonstrates that CII excitation is the dominant cooling mechanism in cold atomic gas. The resulting line at 158 μm is the strongest far-infrared line in the Galaxy (Figure 5.5) and an important diagnostic of the ISM in other galaxies.

Because the collisional timescale is long compared to spontaneous decay, every collisional excitation produces a radiative de-excitation, and the total cooling rate of the cloud is proportional to the number of CII–e collisions. We discuss collisional excitation in detail in the following chapter on HII regions and here simply state a general equation for the cooling rate per unit volume,

$$\text{Cooling} = n_{\text{H}}^2 \Lambda_0 \left(\frac{T_0}{T}\right)^{1/2} e^{-T_0/T} \equiv n_{\text{H}}^2 \Lambda(T), \tag{5.19}$$

where Λ_0 is a constant that depends on the collisional cross-section and energy of the transition, kT_0. The dependence on the square of the hydrogen number density is because the collisional rate depends on the product of the colliders and both the CII and electron number density scale directly with n_H. The temperature dependence of the cooling function, $\Lambda(T)$, includes an exponential Boltzmann factor term and a $T^{-1/2}$ term that is due to the reduced effectiveness of electrostatic attraction with particle speed.

By balancing heating with cooling, we can determine the equilibrium temperature at a given hydrogen density. Heating scales with the strength of the radiation field and can be arbitrarily high. However, cooling is bound by the carbon abundance $[C]/[H] \sim 10^{-4}$. In strong radiation fields, the heating can overwhelm the CII cooling and the gas temperature will rise to the upper $\sim 8000\,K$ levels that are observed in parts of the neutral ISM. At these high values, the long tail of the Maxwellian distribution reaches a point where some hydrogen can be excited. This then opens a new path for cooling through the $n = 2 \rightarrow 1$ (Lyman α) line. The high energy of this transition and the large abundance of hydrogen make this very effective even if only a tiny fraction of hydrogen is excited. The cooling term has the same form as Equation 5.19 but now with $T_0 = 10.2\,eV = 1.2 \times 10^5\,K$ and a different Λ_0 that is about three orders of magnitude larger due to the combination of relative abundances, cross-sections, and escape probability of the Lyman α photon. The total cooling rate over the full range of temperatures at which hydrogen would remain neutral is shown in Figure 5.6. The steep jump at $\sim 10^4\,K$ is known as the Lyman α wall and effectively sets an upper limit to the temperature of the atomic ISM.

Note that heating, as a radiative process, depends linearly on the gas density but cooling, as a collisional process, has a quadratic dependence.

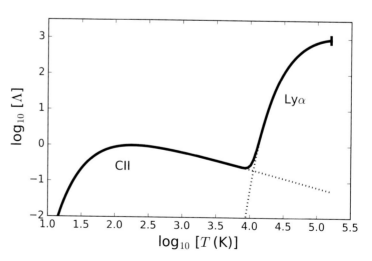

Fig. 5.6. The cooling function, Λ (shown in arbitrary units), as a fuction of temperature for an atomic cloud. At low temperatures, the primary coolant is the fine structure 158 μm line of CII. At high temperatures, $\gtrsim 10^4\,K$, electron collisions excite enough hydrogen to the $n = 2$ level for Lyman α photons to dominate. The vertical line at the end of the curve marks where the thermal energy equals the ionization potential of hydrogen.

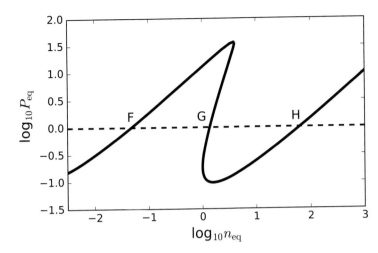

Fig. 5.7. The variation of internal pressure for an atomic cloud in thermal equilibrium as a function of its density (each in arbitrary units). The dashed line shows that the additional constraint of dynamic equilibrium leads to three solutions for the equilibrium density, labeled F, G, and H. F and H correspond to the two-phase neutral medium and G is unstable as small changes in density lead to large jumps in pressure.

The density is often determined through dynamical processes, so for a given equilibrium value, n_{eq}, and fixed radiation field, $\Gamma = \Gamma_0$, the heating and cooling terms balance at a temperature given by the inverse of the cooling function in Figure 5.6,

$$T_{eq} = \Lambda^{-1}\left(\frac{\Gamma_0}{n_{eq}}\right). \tag{5.20}$$

These considerations show that atomic clouds can, in principle, exist over a range of temperatures and densities. However, such clouds would have a range of pressures,

$$P_{eq} = n_{eq}kT_{eq} \propto n_{eq}\Lambda^{-1}\left(\frac{\Gamma_0}{n_{eq}}\right). \tag{5.21}$$

Figure 5.7 plots the equilibrium density and corresponding pressure of a cloud for different temperatures. If we now consider the ISM as a whole, we would expect those clouds with high pressures to push on the clouds with low pressures until a dynamic equilibrium results. The additional constraint of pressure balance, P_{eq} = constant, gives three solutions for n_{eq}. They are labeled F, G, and H after Field, Goldsmith, and Habing who wrote the seminal paper describing these ideas in 1969.

Mathematical solutions do not necessarily indicate a viable physical situation. We must also consider the stability of each solution. Picture a perturbation to this atomic medium, such as a compression wave passing through (possibilities include a supernova or expanding HII region which we will discuss in Chapter 8). If the density of a cloud at points F and H is increased, the condition of thermal equilibrium, i.e., following the curve, would imply that its internal pressure would rise and act against the external compression: the cloud is stable.

At point G, however, an increase in density would lead to a jump to low pressures as if the cloud were to implode. This is clearly not stable.

Thus, we conclude that atomic clouds should lie in one of two stable phases: either low density and therefore warm, or high density and cold. These correspond to states F and H respectively and are called the **warm neutral medium** (WNM) and **cold neutral medium** (CNM). Approximate values for the density and temperature in each phase are $n_H \simeq 10^6, 10^8 \text{ m}^{-3}$ and $T_H \simeq 8000, 100 \text{ K}$ respectively.

This elegant, 50-year old theory was an important step toward understanding the early 21 cm observations. In practice, there are additional heating and cooling terms not discussed here and the timescales of different processes vary, so the premise that the ISM satisfies the dual requirements of both thermal and dynamical equilibrium undoubtedly breaks down in some regions. Nevertheless, the general reasoning and qualitative results from this heuristic picture help us understand how quantum processes shape the structure of the ISM on Galactic scales.

5.5 Gas-to-Dust Ratio

There is a striking similarity between the dust and atomic gas maps in the Galaxy montage in Figure 1.1. This shows that these two components are well mixed in the ISM. Figure 5.8 plots the specific intensity at 857 GHz (350 μm) versus the 21 cm emission point by point in the all-sky maps. Starting from Equation 4.14, we can relate the y-axis in MJy per steradian to the dust surface density, Σ_d,

$$F_\nu/\Omega = I_\nu = B_\nu \tau_\nu = B_\nu \int \rho_d \kappa_\nu^{dust} ds = \kappa_\nu^{dust} B_\nu \Sigma_d, \qquad (5.22)$$

where we have cavalierly assumed constant values in the integration steps here (equivalent to using appropriately weighted means). The figure shows a linear relation, $I_{857GHz}/N_{HI} \simeq 6.5 \text{ MJy/sr}/10^{25} \text{ m}^{-2}$, which implies that the surface densities of dust and gas are proportional to each other. Using values appropriate to interstellar dust (see Chapter 4), $\kappa_{857GHz}^{dust} \approx 0.2 \text{ m}^2 \text{ kg}^{-1}$, $T_d \approx 20 \text{ K}$, this empirical relation between observed quantities translates to a gas-to-dust mass ratio,

$$\frac{M_H}{M_{dust}} \approx 100. \qquad (5.23)$$

The ISM is exceptionally polluted by terrestrial standards. The cities with the worst air quality in the world have densities of micron-sized particulates $\sim 10^{-7} \text{ kg m}^{-3}$, which corresponds to a relatively spotless gas-to-dust mass ratio of 10^7.

Note that although the dust and gas are well mixed in the ISM, they are not generally thermally coupled. The average dust temperature in

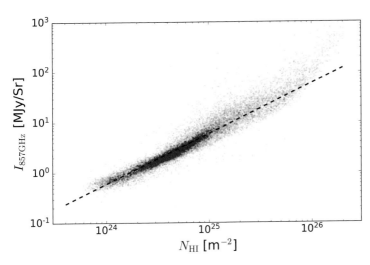

Fig. 5.8. The specific intensity at 857 GHz (350 μm) measured with the Planck satellite versus the HI column density from the HI4PI survey for each pixel in the all-sky maps in Figure 1.1. The dashed line shows a constant ratio, $I_{857\,GHz}/N_H = 6.5\,MJy/sr/10^{25}\,m^{-2}$.

the diffuse ISM is $\sim$19 K (Chapter 4), whereas we have shown here that there are cold and warm atomic clouds with temperatures of $\sim$100 K and $\sim$8000 K respectively. Collisions are too rare at the typical low densities of the atomic (and ionized) clouds for thermal equilibrium to be established so the temperature for each component, dust and gas, is regulated through different absorption and emission processes.

The smoggy interstellar medium blocks starlight and the uniform gas-to-dust ratio manifests itself through a constant gas-to-extinction ratio, with a value at V band of

$$\frac{N_H}{A_V} = 2 \times 10^{25}\,\mathrm{m}^{-2}\,\mathrm{mag}^{-1}. \tag{5.24}$$

There are some variations in this ratio due to differences in the numbers of small dust grains (see Chapter 4) though there is little overall variation on cloud scales or at slightly longer, near-infrared, wavelengths. Its value can be directly measured through observations of hydrogen Lyman α absorption in ultraviolet spectra of hot stars and their reddening. The same technique of measuring ISM absorption lines in stellar spectra allows us to measure abundances of other elements in the gas phase and thereby learn about the composition of the dust.

5.6 Equivalent Width and the Curve of Growth

Just as dust extinguishes starlight, so absorption lines of stellar spectra reveal the gas components of the ISM. Photons with the right energy can excite electronic levels. The energy separations are several eV so hot stars (generally spectral type O) that produce large amounts of

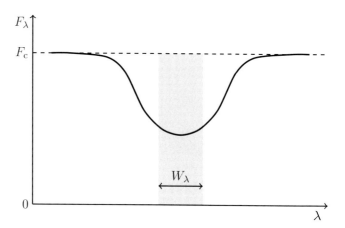

Fig. 5.9. Schematic of an absorption line from a gas-phase species in the ISM seen against a background continuum source. The gray shaded region has the same area as the absorption dip and represents the equivalent width.

ultraviolet radiation are the best background sources for these types of study. The absorption wavelength tells us the element and its ionization state. ISM lines can be distinguished from those in the stellar photosphere as they are of neutral or lightly ionized species indicating low temperatures, and often have different radial velocities from that of the star.

As with the measurement of the HI temperature above, the cloud produces an absorption dip in the continuum of the background sources,

$$F_{\text{obs}}(\lambda) = F_{\text{c}}\, e^{-\tau_\lambda}, \tag{5.25}$$

where the common usage for these ultraviolet observations is to express the intensity and optical depth in terms of wavelength. The total amount of absorption is the area of the dip and is parameterized by the **equivalent width**, W_λ, which is defined relative to the continuum intensity,

$$I_{\text{c}}\, W_\lambda = \int (I_{\text{c}} - I_{\text{obs}})\, d\lambda, \tag{5.26}$$

and illustrated in Figure 5.9. W_λ has units of wavelength, traditionally Å, and can be thought of as representing the wavelength range over which a binary transmission, either purely on or off, would produce the same decrement in the number of received photons. Its historic utility was due to its insensitivity to the spectral resolution of the observations, including unresolved lines. However, reducing a spectrum to a single number generally loses some information such as multiple components along the line of sight, as seen for HI 21 cm absorption (Figure 5.4).

For low optical depths, $\tau_\lambda < 1$, the exponential can be expanded to first order, which leads to the relation,

$$W_\lambda = \int (1 - e^{-\tau_\lambda})\, d\lambda \simeq \int \tau_\lambda\, d\lambda = \iint \kappa_\lambda\, d\lambda\, ds \propto \int n\, ds, \tag{5.27}$$

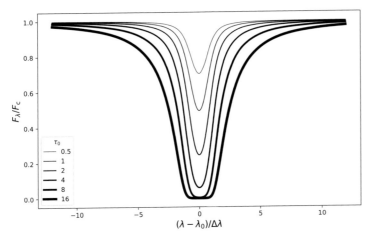

Fig. 5.10. Absorption lines, normalized to the continuum, for different peak optical depths. The line profiles are Voigt with equal FWHM and amplitude for the Gaussian and Lorentz terms.

where n is the number density of absorbers and we use Equations 3.7 and 3.18. In this limit, the equivalent width is proportional to the column density of the species producing the absorption.

As the optical depth approaches unity, the above approximation is no longer valid and we need to numerically integrate over the line function. Figure 5.10 shows that the continuum normalized profile bottoms out at zero corresponding to complete absorption, for $\tau_\lambda \gg 1$ at line center. The equivalent width still increases with central optical depth but more gradually now as the increase is not in the depth of the line but in its width. To quantify this requires a prescription for the wavelength dependence of optical depth.

Doppler broadening by either thermal or large-scale motions is typical for most ISM lines and results in a Gaussian form,

$$\tau_\lambda = \tau_0 e^{-(\lambda - \lambda_0)^2 / 2\sigma_\lambda^2}, \tag{5.28}$$

where λ_0 is the central wavelength of the line, τ_0 is the peak optical depth, and σ_λ is the dispersion which relates to the full width at half maximum (FWHM), $\Delta\lambda = \sqrt{8 \ln 2}\sigma_\lambda$. Optical depths can be very high for electronic transitions in the ultraviolet and, similar to pressure broadening, the continual bombardment of photons affects the lifetime of the upper energy state. This is manifested through an uncertainty in its energy level which is described by a Lorentz profile,

$$\tau_\lambda = \frac{\tau_0}{1 + [(\lambda - \lambda_0)/\gamma]^2}, \tag{5.29}$$

where λ_0, τ_0 describe the central wavelength and peak optical depth as before and γ is the FWHM. This has much broader line wings than a Gaussian but the amplitude is generally much smaller, so it only

Fig. 5.11. The dependence of equivalent width on peak optical depth of a line. The optical depth profile is Voigt with a Lorentz amplitude and FWHM 2% of the Gaussian values. The three different regions of the curve of growth are labeled and illustrated with representative continuum normalized absorption line profiles.

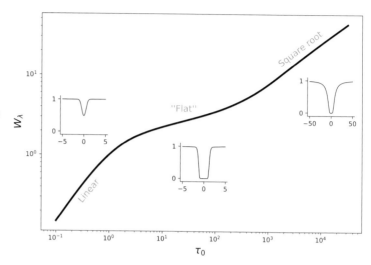

becomes noticeable for extremely high optical depths and therefore column densities. Nevertheless, this is important for measuring the abundances from a few very strong lines in the ISM and of atomic hydrogen at high redshift (Chapter 11). The Voigt profile is a convolution of the Gauassian and Lorentz forms and is plotted in Figure 5.10. This figure exaggerates the Lorentz broadening for illustration. More typically both the Lorentz peak and FWHM are much smaller than the Gaussian term and the wings do not substantially increase W_λ until the combined peak $\tau_0 \gg 10^3$. By integrating model absorption profiles, we can calculate W_λ for a wide range of τ_0. Figure 5.11 plots this function, which is known as the **curve of growth**.

The curve of growth has the linear increase at small τ_0 that we expect from Equation 5.27, then a "flat" region where W_λ increases by only a factor of a few as τ_0 increases by over two orders of magnitude, and finally increases again though more gradually as $W_\lambda \propto \tau_0^{1/2}$, as the Lorentz line wings become prominent at extremely high τ_0.

All absorption lines share this behavior but the particular values of the equivalent width and the optical depth where the Lorentz wings dominate depend on the absorption and emission rates of the transition and are encapsulated through a parameter known as the oscillator strength. From an observational perspective, this means that we can accurately determine column densities from measurements of equivalent width for low or very high optical depths but only poorly in the flat portion of the curve of growth. With this important caveat, we can use absorption line observations to measure elemental abundances in the ISM.

5.7 Elemental Abundances

Elements heavier than hydrogen and helium are somewhat parochially termed **metals** in the astronomical literature. They may exist in different degrees of ionization but all lie in the ground state given the low densities of the ISM. Some species produce optical absorption lines that can be observed from the ground, but most have energy levels that can only be excited by ultraviolet light and require space-based spectroscopy.

Beginning in the 1960s and 70s and continuing to the present day with the Hubble Space Telescope, ultraviolet observations have shown that many elements are far less abundant in the gas than in our Sun. This is not because our Sun has a radically different composition than the ISM but because atoms condense out of the gas and aggregate into the microscopic solid particles, or interstellar dust, that was the subject of the previous chapter. This is evident from the striking pattern of abundance variation with condensation temperature shown in Figure 5.12.

Most metals are produced in stellar interiors and expelled through stellar winds or explosive events (supernovae and neutron star mergers). As the hot gas expands and cools down, the elements progressively drop out in reverse order of their condensation temperature. The inverse of the abundance pattern above therefore tells us about the composition of dust grains. For the most abundant metals, carbon, nitrogen, and oxygen, Figure 5.12 shows that considerable carbon and oxygen are depleted from the gas phase but little nitrogen. In fact there is more carbon in dust, in the form of graphite or large hydrocarbon chains and rings, than in the gas as atoms or ions. Much of the oxygen is locked up in water ice in the

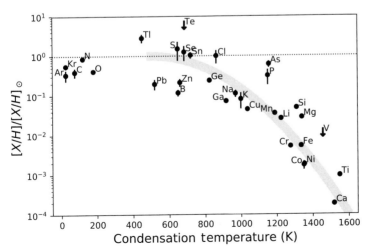

Fig. 5.12. The abundance of elements in interstellar gas relative to solar composition as a function of the element condensation temperature, T_c. Most elements lie below the dashed line indicating depletion in the gas phase. The thick gray line has the form $\log[X/H] \propto T_c^{-2}$, as a guide to the decline in abundance of refractory elements for $T_c \gtrsim 500\,\mathrm{K}$.

cold ISM and in oxides of magnesium, iron, and silicon compounds that are almost entirely depleted from the gas phase. Elements cycle between gas and dust, condensing in the relatively cooler parts and then released back into the gas through grain destruction in interstellar shocks, as part of a Galactic-scale ecology which we explore in Chapter 10.

Notes

Several of the figures in this chapter use publicly available data from the HI4PI survey (HI4PI Collaboration et al., 2016) and the Southern Galactic Plane Survey (McClure-Griffiths et al., 2005). The data points in Figure 5.12 come from a table in Savage and Sembach (1996), who review the determination of elemental abundances in the ISM through ultraviolet absorption lines. Two starting points for the construction and operation of a basic radio telescope for detecting the 21 cm line and mapping the rotation of the galaxy are Parthasarathy et al. (1998) and www.haystack.mit.edu/edu/undergrad/srt/. The two-phase ISM model was introduced by Field et al. (1969) and updated by Wolfire et al. (2003). These show the complexity involved in a full analysis and quantify the schematic plots shown here.

Questions

1. Calculate the collisional timescale for atomic gas with a density $n_H = 10^6 \, m^{-3}$ at 100 K. Compare to the Einstein A value for the 21 cm transition. What density would the gas need to be for radiation to significantly affect the hydrogen level population?

2. Consider a mix of the cold and warm neutral medium along a particular line of sight. Their emission and absorption are blended and you cannot differentiate between the two. If the mix is 50:50, what is the inferred temperature that you measure? What if the cold gas only amounts to 1% of the total column density? Explain why it is challenging to measure the temperature of the WNM.

3a. Create a joint CII-Lyman α cooling function as in Figure 5.6 following the discussion in the associated text. Normalize to the peak of the CII cooling term.

3b. Plot the equilibrium density and equilibrium pressure (in arbitrary units) versus temperature for your cooling function. Plot pressure versus density as in Figure 5.7.

4a. Derive the gas-to-dust mass ratio in Equation 5.23 based on the stated slope of the correlation in Figure 5.8.

4b. Discuss what the existence of such a relation, and its uniformity across the Galaxy, means.

5. Create the curve of growth for a purely Gaussian line profile. Use Equation 5.28, make the wavelength scale dimensionless by setting $\sigma_\lambda = 1$, and calculate W_λ for different τ_0. You should find a similar linear then flat behavior as in Figure 5.11 but not the square root part that comes from Lorentz broadening. For extra credit, find and download code to create a Voigt profile (e.g., astropy.modeling.models.Voigt1D in python) and extend the modeling to incorporate this aspect.

Chapter 6
Ionized Regions

Hot stars produce ultraviolet radiation that can ionize hydrogen and other species. Once ionized, recombination back to the neutral state requires that the ion encounter an electron, which is a slow process in diffuse gas. Consequently much of the Galaxy is filled with a pervasive low-density plasma and bright nebulae, colored by recombination and collisionally excited lines, around young massive stars and old white dwarfs.

6.1 Photoionization and Recombination of Hydrogen

We summarized the Bohr model of the hydrogen atom in Chapter 3 and reasoned that the $n = 2$ level is so high that almost all neutral hydrogen atoms in the ISM lie in the ground state. Consequently, ionization requires the absorption of a photon, γ, with energy $h\nu \geq E_{IP}$,

$$ H + \gamma \rightarrow H^+ + e^-. \tag{6.1} $$

The cross-section for this interaction depends strongly on frequency,

$$ \sigma_{HI}(\nu) \simeq \sigma_{IP} \left(\frac{\nu}{\nu_{IP}} \right)^{-3}, \tag{6.2} $$

where $\sigma_{IP} = 6.3 \times 10^{-22}$ m^2 is the cross-section at the ionization threshold, $\nu_{IP} = 3.29 \times 10^{15}$ Hz ($\lambda_{IP} = 91.2$ nm). The rate of photoionization of a neutral hydrogen cloud depends on the number of photons and their frequency distribution.

The reverse process is the recombination of a free electron with a proton. This two-body process has a rate that depends on the density of electrons and protons, and their relative velocities. The light electrons move much faster than the protons with a velocity distribution that can be characterized as a Maxwellian distribution at temperature T_e. We then parameterize the volumetric recombination rate as $\alpha(T_e)n_p n_e$, which has units of $m^{-3} s^{-1}$. Since the number density of protons, n_p, and electrons, n_e, is the same, we can simplify this to αn_e^2. A recombined electron may land in any of the electronic states, labeled by quantum number n, and the hydrogen atom then radiatively de-excites through the emission of recombination lines. However, if the free electron recombines directly to the ground state, $n = 1$, the resulting photon has enough energy to ionize another hydrogen atom and the number of neutral hydrogen atoms is unchanged. In many astrophysical situations, we therefore only consider recombinations to levels $n \geq 2$, as shown by the subscript 2 in the approximate power law form,

$$\alpha_2 = 2.6 \times 10^{-19} \left(\frac{10^4 \text{ K}}{T_e} \right)^{0.85} m^3 \, s^{-1}. \tag{6.3}$$

We now use these considerations to deduce the effect of massive stars on the ionization structure of the ISM.

6.2 The Strömgren Sphere

6.2.1 A Uniform, Pure Hydrogen Nebula

The most massive stars, with spectral type O, have surface temperatures $T_* > 3 \times 10^4$ K (Table 6.1). Their emission can be approximated as a blackbody, B_λ, with a peak energy, $E \sim 5hc/kT_* = 12.9$ eV, very close to the hydrogen ionization potential. Detailed models of stellar atmospheres provide the surface temperature and production rate of ionizing photons per second, $\dot{N}_{\text{ionize}}$, for stars of different masses. Earlier spectral types have lower numbers, e.g., O3, are hotter, and produce many more ionizing photons.

O stars produce enough ionizing photons to create large ionized, or HII, regions. We first describe their geometry and then the optical and

Table 6.1. O star temperatures, masses, and ionizing photon rates

SpT	T_*(K)	$M_*(M_\odot)$	$\log_{10}(L_*/L_\odot)$	$\log_{10}(\dot{N}_{\text{ionize}})$
O3	44600	58.3	5.83	49.63
O4	43400	46.2	5.68	49.47
O5	41500	37.3	5.51	49.26
O6	38200	31.7	5.30	48.96
O7	35500	26.5	5.10	48.63
O8	33380	22.0	4.90	48.29
O9	31500	18.0	4.72	47.90

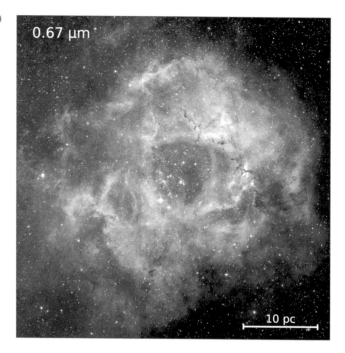

0.67 µm

10 pc

radio light that they produce. To quantify the following calculations, we consider the Rosette nebula, an HII region in Monoceros at a distance of 1600 pc which extends over about a degree in the sky (Figure 6.1). The center of the nebula contains a large stellar group, NGC 2244, that contains four O stars producing a total rate of ionizing photons per second, $\dot{N}_{\text{ionize}} = 10^{49.7}$. A neutral hydrogen atom will have an average ionization cross-section,

$$\langle \sigma_{\text{HI}} \rangle = \frac{\int_{\nu_{\text{IP}}}^{\infty} [B_\nu(T_*)/h\nu] \, \sigma_{\text{HI}}(\nu) \, d\nu}{\int_{\nu_{\text{IP}}}^{\infty} [B_\nu(T_*)/h\nu] \, d\nu} \simeq 3 \times 10^{-22} \, \text{m}^2, \qquad (6.4)$$

where $B_\nu/h\nu$ gives the number of photons at each frequency and we have approximated the radiation field as a blackbody with temperature $T_* = 4 \times 10^4$ K appropriate to the O4V and O5V stars that dominate the ionization.

The ionization rate per neutral hydrogen atom at a radius $r = 1$ pc from the O stars is

$$R_{\text{ionize}} = \frac{\dot{N}_{\text{ionize}}}{4\pi r^2} \times \langle \sigma_{\text{HI}} \rangle \simeq 3 \times 10^{-7} \, \text{s}^{-1}. \qquad (6.5)$$

This is much greater than the recombination rate per hydrogen ion,

$$R_{\text{recombine}} = \alpha_2 n_{\text{e}} \simeq 3 \times 10^{-12} \, \text{s}^{-1}, \qquad (6.6)$$

for typical electron densities and temperatures in the nebula, $n_e \sim 10^7\,\mathrm{m^{-3}}$, $T_e \sim 10^4\,\mathrm{K}$ (values that we will justify from observations later). Expressed in a more relatable way, this implies a neutral hydrogen atom near the center of the nebula would survive, on average, a few months before being ionized but then, once ionized, would have to wait about ten thousand years before recombining with a free electron. If the nebula is in equilibrium, the two states of hydrogen are related by

$$n_{\mathrm{HII}}\, R_{\mathrm{recombine}} = n_{\mathrm{HI}}\, R_{\mathrm{ionize}}, \tag{6.7}$$

and we conclude that, at 1 pc, the nebula is almost fully ionized,

$$\frac{n_{\mathrm{HII}}}{n_{\mathrm{HI}}} \gg 10^4. \tag{6.8}$$

As we move outwards in the nebula away from the O stars, the ionization rate decreases as $1/r^2$ and eventually will match the recombination rate. This defines the boundary of the HII region. The transition from ionized to neutral is the pathlength, l, in the neutral gas corresponding to an optical depth for ionizing radiation, $\tau = n\sigma l = 1$,

$$l = \frac{1}{n_{\mathrm{HI}}\langle\sigma_{\mathrm{HI}}\rangle} \simeq 0.01\left(\frac{10^7\,\mathrm{m^{-3}}}{n_{\mathrm{HI}}}\right)\,\mathrm{pc}, \tag{6.9}$$

where the normalization here is appropriate for the intermediate density neutral material that surrounds a typical HII region. This length scale is over a thousand times smaller than the observed size of the nebula, and is akin to a phase transition between HII and HI.

Stömgren was the first to come to these conclusions in 1939 and the implied radius of the HII region in a uniform medium bears his name. Given that the gas is either almost fully ionized or neutral, we simply equate the rate of ionizations to the rate of recombinations,

$$\dot{N}_{\mathrm{ionize}} = \frac{4}{3}\pi R_{\mathrm{S}}^3\, \alpha_2 n_{\mathrm{e}}^2, \tag{6.10}$$

to give the Strömgren radius,

$$R_{\mathrm{S}} = \left(\frac{3\dot{N}_{\mathrm{ionize}}}{4\pi\alpha_2 n_{\mathrm{e}}^2}\right)^{1/3}. \tag{6.11}$$

For the Rosette, we find $R_{\mathrm{S}} = 40\,\mathrm{pc}$ which is about a factor of 2 larger than its observed extent. The discrepancy is because some of the ionizing photons are absorbed by dust, a correction that we account for later.

The nebula is young enough that it has a higher pressure than its surroundings. As we saw in Chapter 5, the atomic medium consist of two phases, the CNM and WNM, in pressure equilibrium at $P/k = n_{\mathrm{H}}T_{\mathrm{H}} \simeq 10^{10}\,\mathrm{m^{-3}\,K}$. The HII region will therefore expand to lower densities and larger sizes (the dynamics are described in Chapter 8), eventually reaching equilibrium at $n_e \sim 10^6 \, \mathrm{m}^{-3}$ if the temperature remains the same, $T_e \sim 10^4 \, \mathrm{K}$. At this density the Strömgren radius is very large,

$$R_S \simeq 70 \left(\frac{\dot{N}_{\mathrm{ionize}}}{10^{49} \, \mathrm{s}^{-1}} \right)^{1/3} \left(\frac{T_e}{10^4 \, \mathrm{K}} \right)^{0.28} \left(\frac{n_e}{10^6 \, \mathrm{m}^{-3}} \right)^{-2/3} \mathrm{pc}, \quad (6.12)$$

and the total ionized mass is about three orders of magnitude greater than the mass of the star,

$$\begin{aligned}
M_{\mathrm{HII}} &= \frac{4}{3} \pi R_S^3 n_p m_p \\
&= \dot{N}_{\mathrm{ionize}} / \alpha_2 n_e \\
&= 3.2 \times 10^4 \left(\frac{\dot{N}_{\mathrm{ionize}}}{10^{49} \, \mathrm{s}^{-1}} \right) \left(\frac{T_e}{10^4 \, \mathrm{K}} \right)^{0.85} \left(\frac{n_e}{10^6 \, \mathrm{m}^{-3}} \right)^{-1} M_\odot.
\end{aligned}$$
$$(6.13)$$

Such a huge volume is required to balance the low probability of recombination with efficient ionization. The discussion here is highly simplified but demonstrates that O stars have a dramatic effect on the ISM (not even accounting for supernovae, which we return to in Chapter 8). Next, we incorporate a little extra complexity through density variations, consider the effect of dust, and extend the same reasoning to ionization of helium.

6.2.2 The Case of a Density Gradient

The ISM is not homogeneous and stars tend to form in the denser regions of a cloud. We can generalize the Strömgren radius calculation in Equation 6.11 to the case of a radially decreasing density gradient,

$$n_e(r) = n_{e,0} \left(\frac{r}{r_0} \right)^{-p}. \quad (6.14)$$

The ionization–recombination balance is now a sum over contributions in different radial shells,

$$\begin{aligned}
\dot{N}_{\mathrm{ionize}} &= \int_{R_*}^{R_S} 4\pi r^2 \alpha_2 n_e^2 \, dr \\
&= 4\pi \alpha_2 n_{e,0}^2 r_0^{2p} \int_{R_*}^{R_S} r^{2-2p} \, dr \\
&= \frac{4\pi \alpha_2 n_{e,0}^2 r_0^{2p}}{3 - 2p} \left[R_S^{3-2p} - R_*^{3-2p} \right], \quad (6.15)
\end{aligned}$$

for $p \neq 3/2$. Here the lower limit of integration is the stellar radius, $R_* \sim 10^9 \, \mathrm{m}$, which is non-zero so the density does not diverge but is negligible compared to the $\sim 10^{18} \, \mathrm{m}$ scale of R_S. For $p < 3/2$, the exponent $3 - 2p > 0$ and the R_S term dominates in the square brackets leading to a simple analytical solution. For $p > 3/2$, however, the exponent is negative and the integral converges to a finite number as the outer radius

increases. Physically, the density falls off so steeply that the recombination rate decreases faster with radius, $R_{\text{recombine}} \sim r^{-2p}$, than the
increase in volume, $\sim r^3$. Therefore, if $\dot{N}_{\text{ionize}}$ is large enough, the ionization rate is greater than the recombination rate at all radii and the HII
region is unbound. In an inhomogeneous cloud, this naturally leads to a
breakout of the ionized gas on one side, known as a blister HII region.

Ionization is a single-body process that scales linearly with the
hydrogen density, and recombination is a two-body process that varies
as the square of the density. Thus, at sufficiently low densities, ionization will overwhelm recombination. Indeed, by volume, much of the
Galactic ISM is ionized and produces a diffuse glow in hydrogen recombination lines (see below). The most extreme case is the intergalactic
medium, which has extremely low densities, $\ll 10^3 \text{ m}^{-3}$, and is fully
ionized (Chapter 11).

6.2.3 The Effect of Dust

In the above, we have assumed that every ionizing photon from a star
does indeed ionize a hydrogen atom. However, as we saw in Chapter 4
the ISM is very dusty and strongly attenuates radiation. Dust sublimes in
the ISM at a temperature of ~ 1600 K (see Figure 5.12) so its existence
in HII regions (and indeed the WNM) might be surprising. However,
the density of these regions is so low that the gas and dust are thermally
decoupled. Gas heats by energy equipartition with photoelectrons and
dust by direct absorption of starlight. Gas cools through discrete line
emission but dust, as a solid, emits over a continuum and can therefore radiatively cool to much lower temperatures. Consequently dust
survives in HII regions and has a major effect on their size and evolution.

Let's go back to the constant density case but now assume that dust
is uniformly mixed with the gas. We will calculate the size R'_S of a dusty
HII region compared to the dust-free size R_S calculated in Equation
6.11, as a function of the total dust optical depth across the nebula.

We calculate the ionization rate per neutral hydrogen atom at a
distance r from the O stars as in Equation 6.5,

$$R_{\text{ionize}} = \frac{\dot{N}_{\text{ionize}}}{4\pi r^2} e^{-\tau} \langle \sigma_{\text{HI}} \rangle, \tag{6.16}$$

but now we take into account the attenuation of photons through the $e^{-\tau}$
factor. Photons are lost both by dust absorption and hydrogen ionization
and the total optical depth is $\tau = \tau_{\text{HI}} + \tau_{\text{d}}$. The recombination rate per
hydrogen ion remains the same as in Equation 6.6, and in equilibrium
they balance at each radius r,

$$\frac{\dot{N}_{\text{ionize}}}{4\pi r^2} e^{-(\tau_{\text{HI}}+\tau_{\text{d}})} n_{\text{HI}}\sigma_{\text{HI}} = \alpha_2 n_{\text{e}}^2 \tag{6.17}$$

(where we implicitly use averages for the optical depths and cross-sections). We reorganize this by substituting for $\dot{N}_{\text{ionize}}$ from Equation 6.10,

$$e^{-\tau_{\text{HI}}} n_{\text{HI}} \sigma_{\text{HI}} = \frac{3r^2 e^{\tau_d}}{R_S^3}.$$
(6.18)

Recognizing that $d\tau_{\text{HI}} = n_{\text{HI}}\sigma_{\text{HI}} dr$, we can convert this to an integral,

$$\int_0^\infty e^{-\tau_{\text{HI}}} d\tau_{\text{HI}} = \int_0^{x'} 3e^{\tau_d} x^2 dx,$$
(6.19)

where $x = r/R_S$ and the integration limits range from the nebula center to its edge, $x' = R_S'/R_S$. The corresponding HI optical depth ranges from 0 to effectively infinity where the hydrogen becomes fully neutral, so the left hand side of the equation integrates to unity. Because the dust is well mixed with the gas and the density is uniform, its optical depth increases linearly with radius, $\tau_d = x\tau_{dS}$, where τ_{dS} is the optical depth from the star to the edge of a dust-free Strömgren sphere. Integrating by parts, we then find,

$$(\tau_{dS}'^2 - 2\tau_{dS}' + 2)e^{\tau_{dS}'} = 2 + \frac{\tau_{dS}^3}{3},$$
(6.20)

where $\tau_{dS}' = x'\tau_{dS}$ is the dust optical depth from the stellar surface to the edge of the dusty nebula. This can be solved numerically to find the fractional size, x', of the dusty HII region to the dust-free Strömgren sphere as a function of its total optical depth, shown in Figure 6.2. As τ_{dS}' increases, the size of the HII region shrinks but it still remains within $\sim 20\%$ of the dust-free region for optical depths $\tau_{ds} = 1$. The volume, which is a measure of the number of photons that ionize hydrogen, decreases much more rapidly. The dashed line shows the complement,

Fig. 6.2. The ratio of the size of a dusty HII region compared to the dust-free case, as a function of its total dust optical depth. The dashed line shows the ratio of the number of photons absorbed by dust compared to those that ionize hydrogen.

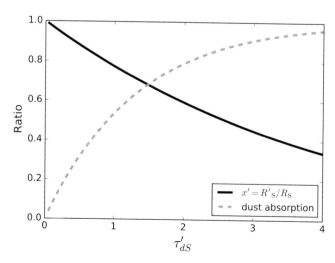

$1 - x'^3$, which is the fraction of ionizing photons that are absorbed by dust. Large HII regions such as the Rosette have relatively low column densities, $\tau_{ds} < 1$ (we can see the central O stars in optical images). However, at early times, they are much denser and $\tau_{ds} \gg 1$. Young, compact HII regions are optically invisible but their warm dust produces strong infrared radiation that can be detected across the Galaxy and in the most distant star-forming galaxies.

6.2.4 HeII Regions

Helium is the second most common element in the Universe and, when ionized, produces an analogous HeII region. The importance of such regions is that their recombination lines can be analyzed to measure the helium abundance and thereby test models of Big Bang nucleosynthesis.

With two protons in the nucleus, the helium ionization potential is 24.6 eV, about twice as high as hydrogen. Only the hottest stars produce large numbers of such energetic photons and therefore have HeII regions. The physical principles that govern the size of the HeII region are similar to those discussed for HII regions above but where we take into account the relative production rate of hydrogen-only ionizing photons $\dot{N}_{ionize}^{H}$ with energies between 13.6 and 24.6 eV with the production rate of helium ionizing photons $\dot{N}_{ionize}^{He}$ with energies greater than 24.6 eV.

As there are about 11 hydrogen atoms for every helium, if $\dot{N}_{ionize}^{H} \gtrsim 11 \dot{N}_{ionize}^{He}$, then all the hydrogen will be ionized, and therefore transparent, within the region where helium can be ionized by the most energetic photons. In this case, the HeII region extends as far as the HII region and stops only when the size is so large that the ionization rate of hydrogen drops to its recombination rate and the gas becomes neutral.

In the opposite case, $\dot{N}_{ionize}^{H} \lesssim 11 \dot{N}_{ionize}^{He}$, there will be some neutral hydrogen atoms that compete for the high-energy photons above 24.6 eV. This limits the size of the HeII region but the HII region can extend further as that is determined by the absolute, not relative, value of $\dot{N}_{ionize}^{H}$.

The ionization structure of a combined H–He gas therefore depends on the spectral shape, or color, of the ionizing source. Main sequence stars with spectral types O6–O3 and Wolf–Rayet stars (massive stars that have blown off their outer hydrogen envelope) have surface temperatures $> 4 \times 10^4$ K and produce enough energetic photons to ionize helium across the full extent of the HII region. O stars with spectral types later than O6, i.e., O9–O7, produce a small HII region surrounded by a HeI shell within the HII region.

Finally, HeII regions may also be found around the white dwarfs that result from the death of low and intermediate mass stars, $\lesssim 8\,M_\odot$. These visually spectacular planetary nebulae are small, $\lesssim 1$ pc, and short-lived as white dwarfs cool below 4×10^4 K in $\sim 10^4$ yr.

6.3 Continuum Emission

The ionized gas emits over a broad range of wavelengths due to the curved paths (and therefore acceleration) of electrons as they move within a sea of ions or are deflected by magnetic fields. Young, dense HII regions produce strong bremsstrahlung radiation and the relativistic particles created by supernovae shock waves produce synchrotron radiation. Both are most commonly observed in the ISM at radio wavelengths (Figure 6.3).

I only state the results pertaining to the SED here and give references for their derivation in the Notes section at the end of the chapter.

6.3.1 Bremsstrahlung

Bremsstrahlung radiation is the emission produced by the thermal motions of the ionized gas. This is also known as free–free emission as each ion–electron interaction represents a change in the energy between a continuum of unbound states. The bremsstrahlung optical depth is

$$\tau_\nu = 3.3 \times 10^{-19} \left(\frac{\nu}{1\,\text{GHz}} \right)^{-2.1} \left(\frac{T_e}{10^4\,\text{K}} \right)^{-1.35} \int n_e^2 dl, \qquad (6.21)$$

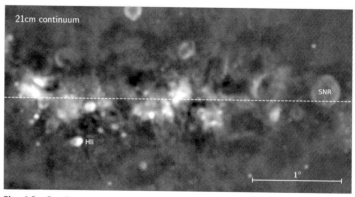

Fig. 6.3. Continuum emission at 1.4 GHz (21 cm) from a $3.3° \times 1.7°$ segment of the Southern Galactic Plane Survey. The dashed line shows Galactic Latitude $b = 0°$. There are numerous thermal (bremsstrahlung) and non-thermal (synchrotron) sources due to HII regions and supernova remnants respectively. The former tend to be bright and compact and the latter often have a shell-like structure, as illustrated by the two labeled cases.

where the integral over the pathlength is known as the **emission measure** (EM) and is proportional to the number of two-body interactions along the line of sight. Because we generally measure number densities in m^{-3} and lengths in pc, EM units are the odd mixture, m^{-6} pc. In these units, EM values range from $\sim 10^{13}$ in low density, old HII regions which have very weak emission, up to $\sim 10^{19}$ in the densest, young HII regions which can be very bright at GHz frequencies. The negative power on the frequency shows that the coupling of the radiation and the plasma is strongest at long wavelengths.

The motions in the plasma are due to thermal energy and the radiation is a blackbody modified by the optical depth as described in Chapter 3. For infrared–radio wavelengths, we can use the Rayleigh–Jeans approximation to find two cases,

$$F_\nu = B_\nu(1 - e^{-\tau_\nu})\Omega \propto \begin{cases} \nu^2 & \tau_\nu > 1, \\ \nu^{-0.1} & \tau_\nu < 1, \end{cases} \quad (6.22)$$

assuming uniform density. The optically thin spectral index is very flat, $\alpha = d\log F_\nu/d\log \nu = -0.1$, over a wide range, until the exponential cutoff in the optical where $h\nu \sim kT_e$. A schematic SED is shown in Figure 6.4 for our fiducial Rosette Strömgren sphere with $n_e \sim 10^7\,m^{-3}, T_e \sim 10^4\,K, R_s = 25$ pc.

In practice, the emission in the infrared is dwarfed by the thermal dust emission at $\lambda \lesssim 1$ mm, but the change in the bremsstrahlung spectral index from optically thick to thin occurs at much longer wavelengths and is an identifiable feature. The turnover frequency at which this happens, ν_1, can be determined quite accurately from multi-wavelength radio observations. The electron temperature can be measured directly

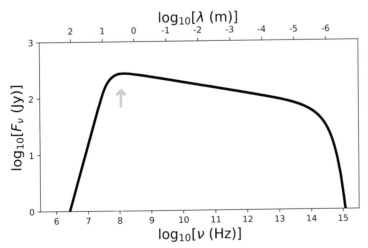

Fig. 6.4. Schematic SED for the bremsstrahlung radiation from an HII region with the average density, temperature, and angular size as the Rosette nebula. The steep slope at low frequencies where the emission is optically thick is equal to the Rayleigh–Jeans slope of two, whereas the optically thin emission emits over a broad fairly flat plateau until the Wien limit at high frequencies. The arrow indicates the turnover frequency.

from resolved long wavelength images in the optically thick part of the SED since the emission is a blackbody, $F_\nu = B_\nu \Omega \propto T$. Then setting $\tau(\nu_1) = 1$ in Equation 6.21 gives the emission measure, EM, and thereby the root mean square of the density, $\langle n_e^2 \rangle^{1/2} = (EM/L)^{1/2}$, where L is the diameter of the HII region. We can then derive the number of ionizing photons from the Strömgren sphere Equation 6.10 and learn about the properties of the central stars.

For the Rosette, observations show $\tau = 1$ at $\nu_1 \sim 500\,\mathrm{MHz}$ equivalent to $\lambda_1 = 0.6\,\mathrm{m}$. The SED is near its peak at this wavelength with a flux density of a few hundred janskys. We noted above that the Rosette is over-pressured with respect to the general ISM. As the HII region expands, the emission measure will decrease. Scaling from ionization–recombination balance,

$$\mathrm{EM} = 2n_e^2 R_S \propto \frac{\dot{N}_{\mathrm{ionize}}}{R_S^2}. \tag{6.23}$$

This implies that the turnover frequency will become lower as the region expands. Note, however, that the flux density in the optically thin regime, $F_\nu = B_\nu \tau_\nu \Omega$, remains the same since $\Omega \propto R_S^2$ (i.e., the larger size compensates for the lower surface brightness). That is, the SED would extend further to the left in Figure 6.4. Conversely, at earlier times, the HII region will be smaller with a higher surface brightness and the SED would turnover at a higher frequency. Figure 6.5 illustrates the expected evolution in the SED of the Rosette from a young, small dense region to an extended, large region at late times, under the constraint that $n_e^2 R_S^3 = \text{constant}$.

We generally don't witness the evolution of an HII region on human timescales but can use the SED to assess the evolutionary state of any

Fig. 6.5. The bremsstrahlung SED for a series of HII regions with different densities, n_e, and Strömgren radii, R_S, satisfying ionization–recombination balance, $n_e^2 R_S^3 = \text{constant}$. The legend shows the values for each SED with the density in m^{-3} and the radius in pc. An HII region would evolve from a dense, compact state (thick line) to a low-density, large region (thin line).

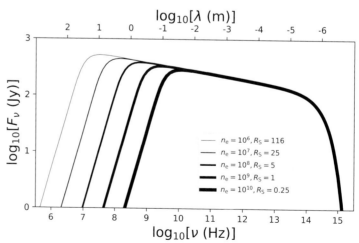

given region. The vertical scale depends on the source distance and ionizing flux but the turnover frequency and size are simple and robust indicators. The youngest, ultra-compact, HII regions have turnover frequencies of several GHz and sizes $\lesssim 0.1$ pc that can be resolved with the Very Large Array in New Mexico. The inferred emission measures are very high, $EM > 10^{19}$ m^{-6} pc, with implied densities $\langle n_e^2 \rangle^{1/2} \gtrsim 10^{10}$ m^{-3}. These are the birthsites of massive stars. Much more massive compact HII regions have even been seen in nearby starburst galaxies where the size and density suggest the formation of super star clusters of $\sim 10^6$ stars (Chapter 11).

6.3.2 Synchrotron

Synchrotron radiation is named after the terrestrial particle accelerators in which it was first seen, but is also observed toward astronomical objects. The emission arises from relativistic particles, mostly protons, gyrating around magnetic field lines and these become the **cosmic rays** that are important for ionizing the dusty, molecular regions of the ISM which we will explore in Chapter 7. Supernovae and Active Galactic Nuclei (AGN) strongly shock their surroundings and produce fast moving charged particles. Scattering off a converging shock front further accelerates them and results in a power law distribution of energies, $N(E) \propto E^{-p}$, a process known as Fermi acceleration.

For optically thin emission,

$$F_\nu \propto B^{(p+1)/2} \nu^{-(p-1)/2}, \tag{6.24}$$

where B is the magnetic field and p is the spectral index of the energy distribution. Cosmic ray measurements show $p \approx 2.5$ so the spectral index is relatively steep, $\alpha = d \log F_\nu / d \log \nu \approx -0.75$ (Figure 6.6). The flatter bremsstrahlung spectrum therefore dominates at shorter (though still radio) wavelengths.

The coupling between matter and radiation is stronger at longer wavelengths and the emission becomes optically thick. Because the electron energy distribution is not Maxwellian, the emission is non-thermal and the SED does not follow the blackbody form but is slightly steeper,

$$F_\nu \propto B^{-1/2} \nu^{5/2}. \tag{6.25}$$

The flux peaks where the SED turns over from thick to thin, i.e., where $\tau = 1$. For AGN, the turnover frequency ranges from $\nu_1 \simeq 0.1$ to 10 GHz ($\lambda_1 \simeq 3-300$ cm). As with bremsstrahlung radiation in HII regions, the turnover frequency has a physical significance that can be exploited, in this case, to determine the magnetic field strength. The

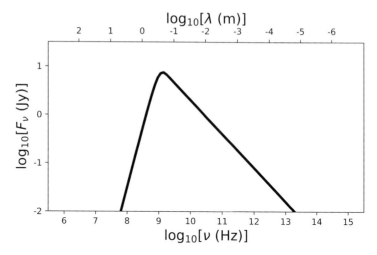

Fig. 6.6. Schematic SED for synchrotron radiation for $p = 2.5$. In contrast to bremmsstrahlung (Figure 6.4), the emission is more strongly peaked with power law indices of 2.5 and 0.7 at low and high frequencies respectively.

spectral slope in the optically thin regime, at higher frequencies, relates to the particle energy distribution.

The synchrotron SED is more sharply peaked than bremsstrahlung but, for a powerful source such as an AGN or starburst galaxy (see Chapter 11), it is the dominant radio component.

6.4 Line Emission

Most of the optical radiation from ionized regions is in line emission. We discuss first the recombination lines of hydrogen and then the collisional lines from trace species that provide key diagnostics of the density and temperature of the gas.

6.4.1 Recombination Lines

The recombination of a free electron with a proton can occur to any of the energy levels n in Figure 3.6. If $n > 1$, the excited hydrogen atom will then decay to lower levels until it reaches the ground state, $n = 1$. This cascade produces a set of photons, the first with an energy corresponding to the potential of the previously unbound electron–proton pair and then the others at fixed values corresponding to the discrete energy jumps between different n in Equation 3.25.

The spectral series that result have an important nomenclature that recognize their respective discoverers. Those most commonly observed in the ISM are Lyman (transitions to $n = 1$), Balmer ($n = 2$), Paschen ($n = 3$), and Brackett ($n = 4$). Thus a recombination of a free electron directly to the ground state produces a Lyman continuum photon and would have an energy greater than $E_1 = 13.6\,\text{eV}$. A recombination to

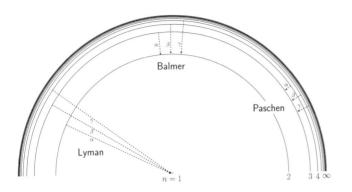

$n = 2$ produces a Balmer continuum photon and so on. The cascade of an excited hydrogen atom between discrete energy levels is labeled by Greek letters, with α representing $\Delta n = 1$, β for $\Delta n = 2$, γ for $\Delta n = 3$, and so on. Thus $n = 2$ to $n = 1$ is called a Lyman α transition and produces a photon with energy $\Delta E = E_2 - E_1 = 10.2$ eV. The transition $n = 3 \rightarrow 1$ is Lyman β, whereas $n = 7 \rightarrow 4$ is Brackett γ. Transitions to $n = 2$ produce Balmer photons which are labeled "H", as in Hα for $n = 3 \rightarrow 2$, Hβ for $n = 4 \rightarrow 2$, and so on. This is schematically shown with wavelengths of various transitions in Figure 6.7.

Following this convention, let's consider the variety of ways in which a hydrogen recombines and decays to the ground state. The most direct path, that of a free electron to $n = 1$, produces a Lyman continuum photon. But such a photon has enough energy to ionize any other neutral hydrogen, and we showed in Equation 6.9 that it does not have to travel very far before this happens. That hydrogen atom will then be ionized, will eventually recombine, and, if it also produces a Lyman continuum photon, the process repeats. This scattering process produces a diffuse ionizing radiation field across the HII region. Consequently, the Strömgren sphere Equation 6.10 balances ionization to recombinations only at levels $n \geq 2$.

Now consider an electron recombining into a higher state, $n > 1$. This produces a (Balmer, Paschen, Brackett, etc.) continuum photon and an excited hydrogen atom that will decay through either a Lyman α, β, γ, etc. line or a higher level series. Once again, if a Lyman photon is produced, it will be readily absorbed by any neutral hydrogen atoms as they will be in the ground state and have a high optical depth. That excited hydrogen atom may decay through the emission of the same Lyman line (in which case the process repeats) or it decays through two or more intermediate steps. Ultimately, all recombinations produce a Balmer photon ($n \geq 3 \rightarrow 2$) and a Lyman α photon. The latter will

Fig. 6.8. Hα map of the Galaxy revealing the presence of a pervasive ionized medium. The map size is −180° to +180° in Galactic Longitude and −60° to +60° in Galactic Latitude (the same as the multi-wavelength panoramas in Figure 1.1). The data are from all-sky maps created by the Wisconsin Hα mapper.

resonantly scatter until either it is absorbed by dust or it decays as two photons through a virtual state between $n = 1$ and 2.

The sequence we have described above can be considered a random walk for any one atom with quantum rates governing the transition between different energy levels. When applied to a macroscopic HII region, the probabilities result in the recombination line strengths being in a fixed proportion. In particular, there is approximately one Hα photon for every two recombinations. Conversely, the observation of any recombination line relates back to the number of ionizations and therefore the properties of the central star. We will see in Chapter 11 how this can be used to make a simple but powerful measure of the star formation rate in other galaxies.

Figure 6.8 shows that the Galaxy glows with widespread Hα emission. The emission is brightest toward relatively young HII regions such as the Rosette discussed above, but there is broad, faint emission from a pervasive low-density **warm ionized medium** (WIM). This is the ionized counterpart to the warm phase of atomic gas (WNM) discussed in Chapter 5, and is produced, and maintained, by the leakage of ionizing photons from HII regions that have burst out of their dense birthplace.

Recombination lines occur at all energy levels. For fixed Δn, the energy difference decreases as n^{-3} and the lines shift to the infrared in the Brackett series and ultimately into the radio regime for $n \gtrsim 60$. Though these lines are much weaker than Hα, they are not affected by dust absorption. Although we have focused on hydrogen here, all ionized species recombine and produce line radiation in the same way. This is an important way in which the helium abundance is measured. However, other species such as oxygen produce much stronger lines through collisional excitation.

6.4.2 Collisional Lines

As noted at the start of this chapter, the energy difference between the ground state of hydrogen and its next level, $n = 1 \rightarrow 2$, is much greater than the kinetic energy of the gas. Collisional excitation of hydrogen, and helium for similar reasons, is therefore very rare. Heavier and rarer elements, however, have additional shells of electrons and more complex energy level diagrams that span the few eV range which can be more readily excited through collisions. Moreover, if ionized, these elements present a large cross-section for interaction with an electron. Consequently, and in spite of their low abundances, the collisionally induced line emission from heavy ions provides much of the optical emission from HII regions.

The ionization potential for oxygen, 13.6 eV, is almost exactly the same as for hydrogen. The same energetic photons that produce an HII region can therefore ionize oxygen. The second ionization potential is much higher, 35.1 eV, but can be achieved around the hottest stars, similar to HeII regions. For illustrative purposes here, we confine the subsequent discussion to lines of singly ionized and doubly ionized oxygen, OII and OIII respectively, but the concepts extend to other species that also exhibit bright lines such as ions of nitrogen and sulfur.

We first describe the basic physics of collisional excitation and de-excitation between two levels and then apply it to model the ratio of two lines from three-level systems.

The Two-Level System

To interpret the spectral lines we observe from the ISM, we need to describe how an electron jumps between quantized energy levels in an atom or, in this case, ion. The simplest example that illustrates the basic principles is a two-level system, Figure 6.9. Collisions can excite the ion, moving it from level 1 to 2, or de-excite it, moving it from level 2 to 1, with rates C_{12}, C_{21} respectively. An excited ion can also spontaneously emit radiation at the Einstein rate, A_{21}. Radiation can also be absorbed, exciting the ion or stimulating emission. The latter

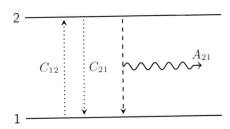

Fig. 6.9. Schematic of collisional excitation and de-excitation and radiative de-excitation for a two-level system.

processes are described by the Einstein B values but we will assume low optical depths and ignore their effects here.

Collisions are a two-body process so we parameterize the rates by the electron density, as they are the primary collisional partner,

$$C_{21} = n_e \gamma_{21},$$ (6.26)

where γ_{21} is the **collisional rate coefficient**. If we consider a system so dense that it is dominated by collisions with no radiative de-excitation, the number of collisions up is balanced by the number down, $n_1 C_{12} = n_2 C_{21}$. In statistical equilibrium, we also know that the level populations are in the ratio $n_2/n_1 = g_2/g_1 \, e^{-E_{12}/kT_e}$, where g_1 and g_2 are the statistical weights at each level. Together, this implies the collisional excitation rate is the same as the de-excitation rate modified by the Boltzmann factor for the associated energy barrier,

$$\gamma_{12} = \left(\frac{g_2}{g_1}\right) \gamma_{21} \, e^{-E_{12}/kT_e}.$$ (6.27)

This relation is derived only in a particular physical situation, that of collisional equilibrium, but it must apply in general because the cross-sections are quantum properties of the ion and independent of environment. This is another example of detailed balance that we introduced Chapter 3.

In the more general situation with radiative de-excitation, the statistical balance between the two levels is

$$n_1 n_e \gamma_{12} = n_2 n_e \gamma_{21} + n_2 A_{21}.$$ (6.28)

The fraction of particles in the upper state is therefore

$$\frac{n_2}{n_1} = \frac{g_2/g_1 e^{-E_{12}/kT_e}}{1 + n_{crit}/n_e},$$ (6.29)

where the **critical density**,

$$n_{crit} = \frac{A_{21}}{\gamma_{21}},$$ (6.30)

relates the relative importance of collisional versus radiative de-excitation. For low densities, $n_e \ll n_{crit}$, collisions are rare, so when the occasional collisional excitation happens, the ion is much more likely to return to level 1 through the emission of a photon than through collisional de-excitation. At the opposite extreme, $n_e \gg n_{crit}$, collisions are common compared to the spontaneous emission rate so most collisionally excited ions return to level 1 through collisional de-excitation rather than radiative decay. Ignoring optical depth effects, the power radiated per unit volume has limiting dependencies,

$$J_{21} = n_2 A_{21} h\nu_{21} \propto \begin{cases} n_e n_1 \gamma_{12} & n_e \ll n_{\text{crit}}, \\ n_1 A_{21} & n_e \gg n_{\text{crit}}. \end{cases} \qquad (6.31)$$

We see the familiar density-squared dependence when collisions dominate the line production, but then saturation to a linear dependence at high density.

Different transitions have different A and γ values and their lines can therefore be used to broadly trace regions of different density in the gas. For many lines, the A values and critical densities are very large compared to typical ISM densities. However, there exist metastable states with low A values that are collisionally de-excited in the terrestrial environment but which produce strong line emission at the low densities of the ISM. These so-called **forbidden lines** are important diagnostics of HII regions.

Of course, the intensity of any observed line also depends on the distance to the source and the element abundance. By extending our description to a third energy level, we can study line ratios that are independent of both of those factors.

The Three-Level System

To simplify our description of a three-level system here, we first restrict the discussion to densities below the critical level for each transition, and then contrast with the high-density limit. Therefore we only need to balance collisional excitations with radiative de-excitation (Figure 6.10).

The statistical balances for levels 2 and 3 are

$$n_1 n_e \gamma_{12} + n_3 A_{32} = n_2 A_{21}, \qquad (6.32)$$

$$n_1 n_e \gamma_{13} + n_2 n_e \gamma_{23} = n_3 A_{31} + n_3 A_{32}. \qquad (6.33)$$

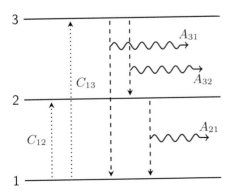

Fig. 6.10. Schematic of collisional excitation and de-excitation and radiative de-excitation for a three-level system.

Since we are in the low-density limit, collisions in the level 2 state are negligible and we can ignore the $n_2 n_e \gamma_{23}$ term. Note that adding the two equations gives the balance for level 1,

$$n_1 n_e (\gamma_{12} + \gamma_{13}) = n_2 A_{21} + n_3 A_{31}. \tag{6.34}$$

We can then solve for the fractional level populations,

$$\frac{n_3}{n_1} = \frac{n_e \gamma_{13}}{A_{31} + A_{32}}, \tag{6.35}$$

$$\frac{n_2}{n_1} = \frac{n_e \gamma_{12}}{A_{21}} + \frac{A_{32}}{A_{21}} \left(\frac{n_e \gamma_{13}}{A_{31} + A_{32}} \right). \tag{6.36}$$

To be more specific, we consider the case of singly ionized oxygen with its energy diagram in Figure 6.11. The levels here are labeled using spectroscopic notation with the superscript referring to the number of spin states, the capital letter S, P, D specifies the orbital angular momentum (0, 1, 2 respectively) and the subscript is the total angular momentum, J. The statistical weight at each level is $2J + 1$. Table 6.2 lists the parameters that we need to calculate the ratios of the two lines.

Fig. 6.11. Energy level diagram for singly ionized oxygen. The first level above ground is split into two fine structure states that emit two lines at nearly the same wavelength, labeled by the numbers in nm.

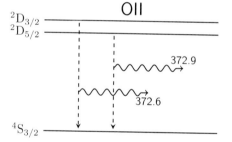

The transition between levels 2 and 3 is in the far-infrared and has a much lower A value than the other two so it does not significantly affect the level populations. The **collision strength**, Ω, is a dimensionless number that packages the quantum properties of each transition in the calculation of the collisional cross-section,

$$\gamma_{21} = \left(\frac{2\pi \hbar^4}{k m_e^3 T_e} \right)^{1/2} \frac{\Omega}{g_2} = \frac{8.63 \times 10^{-12}}{T_e^{1/2}} \frac{\Omega}{g_2} \ \mathrm{m^3 \, s^{-1}}. \tag{6.37}$$

The cross-section for any individual interaction depends on the relative speed of the encounter so the overall rate requires an integration over the Maxwell velocity distribution and brings in a temperature dependence.

The intensity of the line integrated over its frequency profile and direction (Equation 3.13) is $J_{ij} = n_i A_{ij} h \nu_{ij}$ and with $A_{32} \ll A_{31}, A_{32}$ we find

Table 6.2. Transitions of OII

Transition	λ (nm)	A (s^{-1})	Ω
$^2D_{3/2} - {}^4S_{3/2}$	372.6	1.8×10^{-4}	0.54
$^2D_{5/2} - {}^4S_{3/2}$	372.9	3.6×10^{-5}	0.80
$^2D_{3/2} - {}^2D_{5/2}$	...	1.3×10^{-7}	...

$$\frac{J_{372.9}}{J_{372.6}} = \frac{n_2 A_{21} h\nu_{21}}{n_3 A_{31} h\nu_{31}} \simeq \frac{\gamma_{12}\nu_{21}}{\gamma_{13}\nu_{31}} = \frac{\Omega_{21}\, e^{E_{12}/kT_e}\, \nu_{21}}{\Omega_{31}\, e^{E_{13}/kT_e}\, \nu_{31}} \simeq \frac{\Omega_{21}}{\Omega_{31}} = 1.5.$$
(6.38)

The last approximation is because the upper two states are very closely spaced in energy so the exponential factors nearly cancel and the frequency ratio is almost 1. The line ratio is therefore a fixed value independent of temperature in this low-density limit.

The critical density for these transitions is $n_e \sim 10^{8-9}$ m^{-3}, which is in the range of the conditions in the Rosette and other young HII regions. In the high-density limit, the system will be closer to collisional equilibrium and the level populations follow the Boltzmann distribution with a line ratio

$$\frac{J_{372.9}}{J_{372.6}} = \frac{g_2 A_{21} \nu_{21}}{g_3 A_{31} \nu_{31}} \simeq \frac{g_2 A_{21}}{g_3 A_{31}} = 0.30.$$
(6.39)

This is a factor of 5 lower than the low-density limit and also independent of temperature. The line ratio of this system and other species with similar energy level diagrams such as SII are therefore good indicators of the electron density.

Doubly ionized oxygen has a very different energy level diagram, shown in Figure 6.12, and the line ratios consequently exhibit a very different behavior. Note that the ground state is split into three fine structure states that produce far-infrared lines (not shown) so there are two emission lines from level 1 to the nominal ground state. These

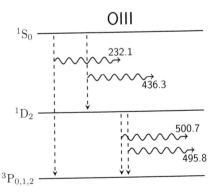

Fig. 6.12. Energy level diagram for doubly ionized oxygen. The bottom level is split into three fine structure states (not shown). The upper two levels are approximately evenly spaced. The numbers refer to the wavelengths of each spectral line in nm.

are so closely spaced in energy that they can be considered as one for understanding the collisional excitation.

Because the two upper levels have a large energy gap, the spontaneous emission rate from level 3 to 2 is high and, unlike OII, must be taken into account in the calculation of the level populations. We consider the line intensity ratios of $3 \rightarrow 2$ to $2 \rightarrow 1$ as these are close in wavelength and can be observed together. This is both more efficient and more accurate as the relative calibration of a single instrument setting is generally more precise than the absolute calibration necessary for comparing line intensities from different instruments or settings. Working from Equations 6.32 and 6.33, you can show that the intensity ratio of the two lines is

$$\frac{J_{21}}{J_{32}} = \left[1 + \left(1 + \frac{A_{31}}{A_{32}} \right) \frac{\gamma_{12}}{\gamma_{13}} \right] \frac{\nu_{21}}{\nu_{32}}. \tag{6.40}$$

The temperature dependence is implicit in the ratio of the collisional cross-sections (see Equation 6.37),

$$\frac{\gamma_{12}}{\gamma_{13}} = \frac{\Omega_{12}}{\Omega_{13}} e^{\Delta E_{32}/kT_{e}}, \tag{6.41}$$

where ΔE_{32} is the (positive) energy difference between levels 2 and 3 ($^1D_2$ and $^1S_0$). If this energy difference is large compared to the thermal energy of the electrons, then the line ratio is strongly dependent on temperature and can therefore be considered a good thermometer of the gas.

The relevant properties for each transition of OIII are listed in Table 6.3. Because $\Delta E_{32}/k \simeq 3 \times 10^4$ K, the relative populations of levels 2 and 3 are very sensitive to variations around the $\sim 10^4$ K nominal temperature of the HII region. The intensity ratio of the doublet 507.7 and 495.8 nm lines from level 2 to 1, relative to the 436.3 nm line from the level 3 to 2, is plotted in Figure 6.13. The sum of the lower level lines is two to three orders of magnitude stronger than the upper level line for $T_e < 10^4$ K. High signal-to-noise spectra are therefore required to detect the 436.3 nm line, which remains relatively weak even in the hottest regions as found in planetary nebulae or around AGN.

The analysis of this line ratio and of other ions with similar level arrangements such as NII shows that HII regions around O stars have

Table 6.3. Transitions of OIII

Transition	λ (nm)	A (s^{-1})	Ω
$^1S_0 - {}^3P_1$	232.1	2.2×10^{-1}	0.28
$^1S_0 - {}^1D_2$	436.3	1.8	0.62
$^1D_2 - {}^3P_1$	495.8	6.7×10^{-3}	2.17
$^1D_2 - {}^3P_2$	500.7	2.0×10^{-2}	...

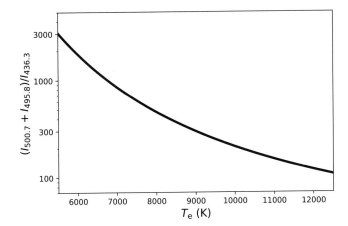

Fig. 6.13. The intensity ratio of the doublet 500.7 and 495.8 nm lines to the upper transition 436.3 nm line in OIII. The lower level lines are much stronger at all temperatures but the ratio decreases with temperature as the greater collisional energy of the electrons increasingly populates the higher levels.

temperatures $T_e \simeq 10^4$K with remarkably little variation from one source to another. The uniformity is because hotter gas produces more numerous and more energetic collisions with an accompanying increase in energy lost from the system through radiation.

6.5 Pulsar Dispersion

Electromagnetic waves interact with electrons even when they are unbound to a nucleus. A completely different way to measure the properties of the ionized ISM is to use the fact that it has a frequency-dependent index of refraction,

$$m = \left[1 - \left(\frac{\nu_p}{\nu}\right)^2\right]^{1/2}, \tag{6.42}$$

where the plasma frequency

$$\nu_p = \frac{1}{2\pi}\left(\frac{e^2 n_e}{\epsilon_0 m_e}\right)^{1/2}. \tag{6.43}$$

Here ϵ_0 is the electric constant and m_e, e are the electron mass and charge respectively. For typical densities in the WIM of $n_e \simeq 10^4\,\text{m}^{-3}$, this is extremely low, $\nu_p \sim 1$ kHz. The corresponding wavelength, $\lambda = 3$ km, is well below what we can observe with radio telescopes (which is limited by the plasma frequency in the much denser ionosphere) so $\nu \gg \nu_p$, which implies $m \simeq 1$ so the effect on radiation is very small. Nevertheless, the vast range of astrophysical phenomena includes sources that are ideally suited to measure it.

Pulsars emit short, intense bursts of radio continuum emission with typical intervals of about a second. These pulses are extremely regular and can be measured with exquisite accuracy. During the passage of a

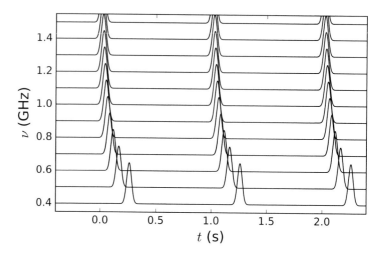

Fig. 6.14. The delay in arrival times of the radio bursts from a pulsar due to its passage through the WIM, here modeled with a dispersion measure, $DM = 10^7 \, m^{-3}$ pc. The pulses arrive at regular intervals at any given frequency but the lower frequencies arrive later than the higher frequencies.

radio pulse through the ISM, the lower frequencies are slowed down slightly more than the higher frequencies. The propagation speed, $v = mc$, implies that the time for the pulse to arrive at the telescope is

$$t = \int_0^d \frac{dl}{v} \simeq \int_0^d \left(1 + \frac{v_p^2}{2v^2}\right)\frac{dl}{c} = \frac{d}{c} + \Delta t, \qquad (6.44)$$

where d is the distance and the time delay

$$\Delta t = \left(\frac{e^2}{8\pi^2\epsilon_0 m_e c}\right)\frac{1}{v^2}\int_0^d n_e dl. \qquad (6.45)$$

We define the **dispersion measure** (DM) as the integral of the electron density along the line of sight,

$$DM = \int_0^d n_e dl. \qquad (6.46)$$

From measurements of the difference in arrival times at different frequencies, we can then derive DM (Figure 6.14). Very long baseline interferometry provides the astrometric accuracy to measure pulsar parallaxes and determine their distance. The average electron density along the line of sight is then $\langle n_e \rangle = DM/d$.

6.6 Heating and Cooling

As with the thermal balance of atomic regions discussed in the previous chapter, so the temperature of HII regions is determined by a balance between heating by fast-moving photoelectrons and cooling through the escape of radiation. In this case, the photoelectrons come directly from the ionization of hydrogen and the cooling is via forbidden line emission predominantly of oxygen.

A photon with frequency $\nu > \nu_{IP}$ can ionize a neutral hydrogen atom and produce an electron with energy $E_{pe} = h(\nu - \nu_{IP})$. If we represent the stellar emission as a blackbody, the average energy of a photoelectron is

$$\langle E_{pe} \rangle = \frac{\int_{\nu_{IP}}^{\infty} [B_\nu(T_*)/h\nu] E_{pe} \, d\nu}{\int_{\nu_{IP}}^{\infty} [B_\nu(T_*)/h\nu] \, d\nu} \simeq 5 \, \text{eV}, \qquad (6.47)$$

as in Equation 6.4. The value here is calculated for $T_* = 4 \times 10^4$ K but changes by less than a factor of 2 from $3 - 5 \times 10^4$ K. The total energy input rate into the HII region is $\dot{N}_{\text{ionize}} \langle E_{pe} \rangle$. Using Equation 6.10, this translates to a volumetric heating rate,

$$\text{Heating} = \alpha_2 n_e^2 \langle E_{pe} \rangle \equiv n_e^2 \Gamma(T). \qquad (6.48)$$

Recombination lines radiate away the residual kinetic energy of the photoelectrons but, because recombination is so much slower than ionization, the recombining electrons will have equilibrated to the gas temperature, $T_e \simeq 10^4$ K, and therefore have an energy $kT_e \simeq 0.9$ eV which is relatively small and can be accounted for by reducing $\langle E_{pe} \rangle \simeq 4$ eV.

Thermal bremmstrahlung emission is also a minor component in the energy loss term as the radiation peaks in the radio regime. The cooling is in fact dominated by collisionally excited forbidden lines of oxygen. These are relatively abundant species that are easily excited and copiously radiate away energetic optical and infrared radiation. For a three-level system described above, the volumetric rate of energy radiated away through the $3 \rightarrow 1$ transition is

$$E_{31} = n_3 A_{31} h\nu_{31} = n_1 n_e \frac{A_{31}}{A_{31} + A_{32}} \gamma_{13} h\nu_{31}, \qquad (6.49)$$

where the second step uses Equation 6.35. Similar terms apply for the corresponding E_{32} and E_{21} terms. Each has the same factor $n_1 n_e = X n_e^2$ where $X = n_1/n_e$ is the abundance of the radiating ion in the ground state. The overall volumetric cooling rate has the generic form

$$\text{Cooling} = n_e^2 \sum_{\text{ions}} X_{\text{ion}} \sum_{\text{transitions}} f(A, \gamma) h\nu \equiv n_e^2 \Lambda(T), \qquad (6.50)$$

where f is a different function for each transition and Λ combines them into a single function of temperature.

We have ignored helium ionization and the radiative transfer of the line but these have relatively small effects in most typical Galactic HII regions. With that caveat, we see that the heating and cooling terms have the same n_e^2 dependence and therefore conclude that the thermal balance is independent of density. Unlike the case for atomic regions, thermal

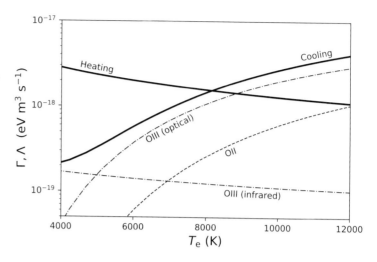

Fig. 6.15. Heating and cooling terms (volumetric rates divided by n_e^2) as a function of the electron temperature in an HII region. The cooling is shown for oxygen equally split between singly and doubly ionized forms, OII and OIII.

equilibrium depends only on temperature through the Γ and Λ terms. These are plotted for prominent oxygen lines in Figure 6.15.

The three dotted and dashed lines that add together to the total cooling term are for singly and doubly ionized oxygen. The structure for OIII is considered as two independent three-level systems, one with the optical lines in Figure 6.12 and the second for the fine structure in the ^{3}P ground state that produces infrared lines at 30–90 μm.

The heating (per unit volume) decreases slowly with temperature as the recombination cross-section is smaller for faster electrons. The fine structure levels in the ground state of OIII are readily populated in the temperature range here and their energy loss term is fairly constant. The higher energy levels of OII and OIII that produce optical emission are only populated by energetic collisions and the loss term is therefore a strongly rising function of temperature. The equilibrium value, where $\Gamma = \Lambda$, occurs at $T_e \simeq 8000$ K. This is clearly stable as a slight rise in temperature leads to reduced heating and enhanced cooling, and vice versa.

It is a coincidence that HII regions have similar equilibrium temperatures as the warm component of HI, but it means that, if the broadly distributed WIM and WNM are in approximate pressure balance, their densities are related by $n_e(\text{WIM}) \sim 0.5 n_H(\text{CNM})$ where the factor of a half accounts for the doubling of the number of particles in the ionized gas.

Notes

Because of their strong line and continuum emission over a wide range of wavelengths, ionized nebulae are some of the most commonly observed objects and the astronomical literature is vast. The primary reference and starting point for many researchers is the textbook by Osterbrock and Ferland (2006). A derivation of the SEDs for bremsstrahlung and synchrotron can be found in the textbook by Rybicki and Lightman (1986). Table 6.1 is a distillation of stellar models published in Martins et al. (2005).

Questions

1a. The most massive and luminous group of stars in the Galaxy is the Arches cluster. It contains about 150 OB stars that produce a total ionizing luminosity $\dot{N} = 10^{51}\,\mathrm{s}^{-1}$. Calculate the Strömgren radius assuming a uniform, dust-free gas with density $n_{\mathrm{H}} = 10^6\,\mathrm{m}^{-3}$.

1b. Calculate the radius assuming dust-free gas with a density dependence $n_{\mathrm{H}} = 10^6 (r/1\,\mathrm{pc})^{-1}\,\mathrm{m}^{-3}$.

1c. Calculate the radius including dust with an optical depth from center to edge of unity (i.e., $\tau'_{\mathrm{dS}} = 1$).

2a. Using the expression for the optical depth in Equation 6.21, show that the bremsstrahlung flux of an HII region is proportional to the number of ionizing photons of the central object. Assume uniform density for simplicity and that the emission is measured at high enough frequencies for the emission to be optically thin.

2b. You observe an HII region at a distance of 1 kpc with a known temperature of 10^4 K. From its radio SED, you find that it turns over at 10 GHz with a flux of 10 Jy. Using the result from above, determine the number of ionizing photons from the central source. Referring to Table 6.1, what stellar type does this correspond to?

2c. From the turnover frequency calculate the emission measure, and thence the radius of the region. What angular size does this correspond to and what telescope could resolve it at the observing frequency?

3. The doublet OIII lines from the $^1D_2$ to ^3P level were considered as a single transition in the discussion above as their energies are so close and they are equally collisionally exicted. In practice, observations at high spectral resolution can differentiate them at 495.8 to 500.7 nm. What is their expected line ratio?

4a. Calculate the time delay of a radio wave at frequency $\nu = 1$ GHz through 1 kpc of ionized gas with density $n_e = 10^4\,\mathrm{m}^{-3}$.

4b. Consider the observation of a pulsar with a finite bandwidth, $\Delta\nu \ll \nu$. What is the time duration over which each radio burst is dispersed?

Chapter 7
Molecular Regions

In dense regions, and away from energetic sources of radiation, atoms can combine to form molecules. The electrons now share orbitals around two or more nuclei with transitions that produce ultraviolet and optical lines analogous to those in atoms. In addition, the interactions of the nuclei themselves have quantized vibrational and rotational states, though at much lower energies. The corresponding lines lie in the infrared and millimeter wavelength regime and pass through the large amounts of dust that generally go hand in hand with molecular regions. Observations reveal a rich spectrum from multiple species that tells us about the physical and chemical properties of the molecular interstellar medium; the coldest parts of the Universe and the sites of stellar birth.

7.1 Molecular Transitions

It is instructive to consider simple models and use classical physics to estimate the magnitude of the energy of molecular transitions in relation to the atomic electronic states that we have discussed in the previous chapters.

We first start with a single atom, X. The electrostatic potential energy, E_{el}, of an electron orbiting the nucleus is proportional to the product of the electron and nuclear charge and divided by the distance between them. The Bohr model gives a characteristic radius,

$$a_0 = \frac{4\pi \epsilon_0 \hbar^2}{m_e e^2}. \tag{7.1}$$

An order of magnitude estimate then gives

$$E_{el} \sim K e^2/a_0 = \hbar^2/m_e a_0^2, \tag{7.2}$$

where $K = 1/4\pi\epsilon_0$ is the Coulomb constant.

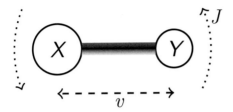

Fig. 7.1. Schematic of the vibrational and rotational modes of a diatomic molecule, with quantum numbers v and J respectively.

We now pair X with a second atom, Y, to form a diatomic molecule, XY. The outer electrons are shared, forming a bond between the nuclei, and lie in quantized orbitals with energy levels of order E_{el}. There are, however, additional degrees of freedom in the system through the vibrational and rotational motions of the nuclei, as schematically shown in Figure 7.1.

The characteristic vibrational energy is $\hbar\omega_{vib}$, where ω_{vib} is the vibrational frequency. The kinetic energy associated with the motion of the nuclei with mass M over scales a is then $Ma^2\omega_{vib}^2$. We can equate this to E_{el} for $a = a_0$ as it would involve the motion of a charge in the electric potential of the system (typical vibrational modes occur on much smaller scales). This reasoning relates $\omega_{vib} \sim \hbar/(m_e M)^{1/2}a_0^2$ and implies

$$E_{vib} = \hbar\omega_{vib} \sim \left(\frac{m_e}{M}\right)^{1/2} E_{el}. \tag{7.3}$$

The rotational frequency, ω_{rot}, has an associated angular momentum, $Ma_0^2\omega_{rot}$, which is quantized in units of $\hbar$. We therefore derive an order of magnitude estimate, $\omega_{rot} \sim \hbar/Ma_0^2$, which implies a characteristic rotational energy,

$$E_{rot} = \hbar\omega_{rot} \sim \left(\frac{m_e}{M}\right) E_{el}. \tag{7.4}$$

Since $M \sim 10^4 m_e$, the relative strengths of electronic, vibrational, and rotational transitions are $1 : 10^{-2} : 10^{-4}$. Typical values are $\sim 10\,eV$, $0.1\,eV$, and $0.001\,eV$, respectively, which correspond to wavelengths $\sim 100\,nm$, $10\,\mu m$, and $1\,mm$. That is, electronic transitions are in the optical/UV, vibrational in the near/mid-infrared, and rotational in the (sub-)millimeter.

7.2 Rotational and Vibrational Lines

The energy levels are determined by solving the Schrödinger equation. Most molecules in the ISM are diatomic and the vibrational and rotational levels are quantized by two numbers, v and J respectively. As the two modes are so widely separated in energy from each other, the wave function can be factored into separate parts. This is known as the Born–Oppenheimer approximation and it means that the vibrational and rotational energy levels are almost fully independent of each other,

$$E(v, J) = E_{\text{vib}} + E_{\text{rot}}. \tag{7.5}$$

There can also be a separate electronic term due to the arrangement of the electron shells which we will come back to later. Here, we restrict our attention to vibration and rotation modes that are excited in the cold and dusty regions where most molecular line observations are made.

The vibrational energies are linearly spaced,

$$E_{\text{vib}} = \left(v + \frac{1}{2}\right) h\omega_{\text{vib}}, \tag{7.6}$$

where ω_{vib} is the vibrational constant. Classically, this can be modeled as a harmonic oscillator with a spring constant relating to the bond strength. Typical ω_{vib} values for diatomic molecules are several 10^{13} Hz. Note that the ground state energy, $v = 0$, is greater than zero. This is known as the zero point energy and is a consequence of the uncertainty principle which implies that the two nuclei can never be completely at rest with respect to each other.

The rotational energies have a quadratic form,

$$E_{\text{rot}} = hBJ(J + 1), \tag{7.7}$$

where B is the **rotational constant** and has units of Hz. Classically, this can be viewed as a rotating, and therefore accelerating, charge. From Equation 7.4, $B \sim 1/Ma_0^2$ so heavier molecules generally rotate slower and have lower rotational energy levels. Typical B values for abundant molecules in the ISM are several 10^{10} Hz. Note that the ground state is at zero so molecules can stop rotating, although they will still be vibrating.

Fig. 7.2. The rotational and vibrational energy levels for carbon monoxide. The left side shows the vibrational energy for each level v. The rotational transitions are illustrated by the gray shading at each level. The rotational energies are about 100 times smaller than the vibrational and the inset on the right hand side shows a zoomed-in region of the J-ladder.

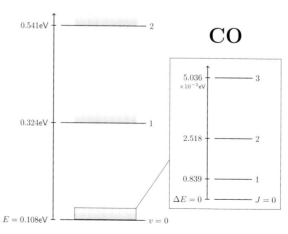

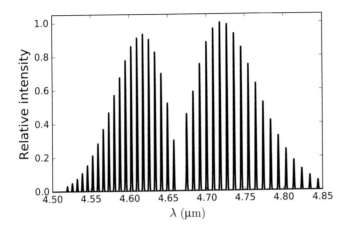

Fig. 7.3. Model spectrum of ro-vibrational lines for CO $v = 1-0$, illustrating the two branches corresponding to a positive or negative change in J and a central gap at $\Delta J = 0$.

Figure 7.2 plots the energy levels for carbon monoxide, $^{12}C^{16}O$ (hereafter CO). This is the most frequently observed molecule in the ISM. Each vibrational level has its own rotational ladder and transitions are from $(v, J) \rightarrow (v', J')$. Certain transitions are much more likely than others due to the similarity between the wavefront solutions for the start and end states. Such **selection rules** show that neighboring pairs are preferred, $\Delta v = \pm 1, \Delta J = \pm 1$. This is a more stringent criterion for the latter case, i.e., vibrational transitions can occur between more widely spaced levels but this is very rare for rotations.

A change in vibrational state can be accompanied by a change in many pairs of rotational states. This produces a multi-lined **ro-vibrational spectrum**, and a simple model for CO $v = 1-0$ is shown in Figure 7.3. The symmetry comes from the sign of the $\Delta J = \pm 1$ jump and produces two branches in the spectrum. The R branch corresponds to a higher energy jump, $J \rightarrow J - 1$, and lies at shorter wavelengths. The P branch is a smaller jump, $J \rightarrow J + 1$, and is at longer wavelengths. The envelope shape arises from the population level distribution that is small at low levels due to the degeneracy, $g_J = 2J + 1$, and at high levels due to the Boltzmann exponential, $e^{E/kT_{ex}}$. The difference between the relative intensity of the P and R branches is due to different values in the Einstein A coefficient. This emission spectrum shown here requires gas at several thousand kelvin for collisions to excite the vibrational levels. Alternatively an absorption spectrum can be detected in colder gas against a bright mid-infrared source, such as an embedded protostar.

The extra bonds and degrees of freedom in molecules with three or more atoms allow many more transitions. This requires additional quantum numbers to describe the vibrational modes and axes of rotation, and different selection effects. Possibilities include a Q-branch with

$\Delta J = 0$. Water, for example, has thousands of lines in the infrared–millimeter region. The range of transitions provides many ways for the molecular ISM to radiate and efficiently cool across a temperature continuum from $> 10^3$ K to ~ 10 K. This has the remarkable consequence that the most common molecule becomes effectively undetectable.

7.3 The Invisibility of H_2 in the Cold ISM

Hydrogen is, by far, the most common element in the Universe and molecular hydrogen is the most common molecule in the ISM. Its symmetry, however, precludes pure rotational transitions. From a quantum standpoint, the two hydrogen atoms are identical so there is no change in state in a $180°$ rotation. Because there is no separation of charge from the center of the system, it is also said to have zero **dipole moment**.

The hydrogen molecule has vibrational modes that can be excited through collisions in shocks at speeds $\gtrsim 20$ km s^{-1} (Chapter 8). For example, the 2.12 µm line corresponds to $v = 1 - 0, J = 3 - 1$ and is an important observational diagnostic of protostellar outflows as it lies in the K-band near-infrared atmospheric window. There are also quadrupole rotational transitions ($\Delta J = \pm 2$ with very small, but non-zero Einstein A values). The lowest energy line is at 28.22 µm from $v = 0, J = 2-0$, which is excited in warm gas, $T > hc/\lambda \sim 500$ K. In colder regions, however, almost all the H_2 will be in the ground state, $v = 0, J = 0$. It will not radiate and is effectively invisible. To diagnose the properties of these regions requires observations of other constituents: dust (Chapter 4) and molecules such as CO with populated rotational levels above the vibrational ground state.

7.4 Tracers of Cold Molecular Gas

The offset between the charge distribution and center of mass in asymmetric molecules such as CO produces a dipole moment and a series of rotational energy levels that can be populated through collisions in cold gas. Although the abundances of these molecules are very low relative to H_2, they provide the only means for the gas to radiate and result in a rich line spectrum at millimeter wavelengths.

Table 7.1 lists a small set of commonly observed, low-lying, rotational transitions, $J + 1 \rightarrow J$, in the ground vibrational level, $v = 0$. Together with the line frequencies, the table shows the energy of the upper state, spontaneous emission rate, and its ratio with the H_2 collisional de-excitation rate (i.e., the critical density) at 10 K.

The Einstein A values are universally extremely small compared to (permitted) vibrational and electronic transitions, but increase higher

Table 7.1. A sample of molecular rotational transitions

Molecule	Transition	ν (GHz)	E_u/k (K)	A (s^{-1})	n_{crit} (m^{-3})
CO	1–0	115.271	5.5	7.20×10^{-8}	2.2×10^{9}
	2–1	230.538	16.6	6.91×10^{-7}	2.3×10^{10}
	3–2	345.796	33.2	2.50×10^{-6}	3.5×10^{10}
^{13}CO	1–0	110.201	5.3	6.29×10^{-8}	1.9×10^{9}
	2–1	220.399	15.9	6.03×10^{-7}	2.0×10^{10}
	3–2	330.588	31.7	2.18×10^{-6}	3.1×10^{10}
CS	1–0	48.991	2.4	1.75×10^{-6}	5.0×10^{10}
	2–1	97.981	7.1	1.68×10^{-5}	8.2×10^{11}
	3–2	146.969	14.1	6.07×10^{-5}	1.2×10^{12}
HCN	1–0	88.633	4.3	2.41×10^{-5}	1.0×10^{12}
	2–1	177.261	12.8	2.31×10^{-4}	9.6×10^{12}
	3–2	265.886	25.5	8.36×10^{-4}	3.7×10^{13}
HCO$^+$	1–0	89.188	4.3	4.25×10^{-5}	1.6×10^{11}
	2–1	178.375	12.9	4.08×10^{-4}	2.9×10^{12}
	3–2	267.558	25.7	1.48×10^{-3}	3.9×10^{12}

up the rotational ladder. The critical densities are similar to the optical forbidden lines of HII regions and are well matched to the conditions of the molecular ISM. Higher transitions are excited by slightly warmer and denser gas.

In addition to CO, the table includes its **isotopologue**, ^{13}CO. An isotopologue is a molecule that consists of at least one less abundant isotope of its constituent elements. They have the same transitions at nearby frequencies with similar decay and excitation rates. The main difference is in their abundance and observations of the rarer species help diagnose conditions in dense regions where lines from the primary species are optically thick. Because the A values, and critical densities, for CO and ^{13}CO are relatively low, they are good tracers of the bulk of the molecular gas, whereas CS, HCN, and other molecules generally locate density enhancements associated with star formation.

Even though ultraviolet and optical light does not penetrate deeply into dusty molecular regions, there are many ionic molecular species such as HCO$^+$. The ionization is caused by cosmic rays, i.e., energetic protons and atomic nuclei, which can penetrate well beyond where optical light is extinguished. Molecular ions play important roles in dynamics as, unlike the rest of the neutral gas, they feel a force as they move relative to magnetic fields. They are also the starting point for much of the chemistry in the gas, as we will explore later.

Well over 100 different molecules, and many of their isotopologues, have been detected in the ISM. The spectra from the dense envelopes

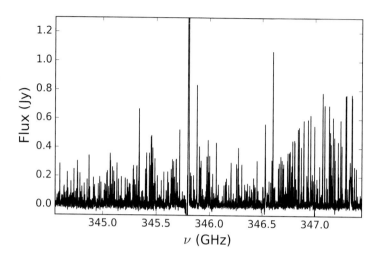

Fig. 7.4. A forest of molecular rotational lines toward the dense gas surrounding the young protostar IRAS 16293-2422. The strongest line at 345.8 GHz is CO $J = 3-2$. This is a small portion of a large spectral line survey in this region observed with the Atacama Large Millimeter Array.

around protostars can consequently become very crowded in the (sub-)millimeter wavelength range, as demonstrated in Figure 7.4.

7.5 Column Density

The strength of a line depends on the number of molecules along the line of sight, the level population distribution, and the transition probability. That is, the column density, excitation temperature, and Einstein A value. Specifically, the form of the absorption coefficient in Equation 3.18 and the relations between Einstein coefficients in Equation 3.15 provide a prescription for the optical depth of a transition at frequency ν from level 2 to 1,

$$\tau_\nu = \int \kappa_\nu ds = \frac{A_{21} c^2}{8\pi \nu^2} \left(1 - e^{-h\nu/kT_{ex}}\right) \frac{g_2}{g_1} \left[\int n_1 ds\right] \phi(\nu). \quad (7.8)$$

The term in square brackets is the column density of the lower state. We can relate this to the upper state through the Boltzmann distribution and integrate over the line profile to derive

$$N_2 = \int n_2 ds = \frac{8\pi \nu^2}{A_{21} c^2} \left(e^{h\nu/kT_{ex}} - 1\right)^{-1} \int \tau_\nu d\nu. \quad (7.9)$$

The optical depth is determined through observations of the specific intensity. Avoiding any background sources, we rearrange the general solution in Equation 3.12,

$$I_\nu = B_\nu(T_{ex}) \left(1 - e^{-\tau_\nu}\right) = \beta B_\nu(T_{ex})\tau_\nu, \quad (7.10)$$

where

$$\beta = \left(1 - e^{-\tau_\nu}\right)/\tau_\nu. \quad (7.11)$$

Molecular line observations of CO and other abundant molecules are often optically thick and the use of β provides a simple a posteriori correction to the column density derived for an optically thin case. Expressing the specific intensity in terms of the brightness temperature, defined in Equation 5.8, gives

$$T_B = \frac{h\nu}{k} \frac{\beta \tau_\nu}{e^{h\nu/kT_{ex}} - 1}. \tag{7.12}$$

Under the approximation of a constant β, this then directly relates column density to the spectrally integrated brightness temperature,

$$N_2 = \frac{1}{\beta} \frac{8\pi k\nu^2}{A_{21}hc^3} \int T_B d\nu, \tag{7.13}$$

where $d\nu = \nu dv/c$ translates the frequency axis to velocity.

The above is similar to the derivation of HI column density from the 21 cm line in Chapter 5 but without using the Rayleigh–Jeans limit, which is no longer valid because the photon energy is larger and the thermal energy smaller in this case. Furthermore, the conversion to a column density necessitates summing over a wider level population,

$$N_{tot} = \sum_i N_i = N_0 \sum_i g_i e^{-E_i/kT_{ex}} \equiv N_0 Q, \tag{7.14}$$

where N_0 is the column density of the ground state, E_i is the energy of level i above the ground state, and we have assumed LTE such that a single excitation temperature describes the full level population. This is usually valid for densities above the critical values of the relevant levels. Q is known as the **partition function** and provides a scale factor for converting a column density measurement at any level, N_i, to the total,

$$N_{tot} = \frac{N_0}{N_i} N_i Q = \frac{e^{E_i/kT_{ex}}}{g_i} N_i Q. \tag{7.15}$$

All the above is general for any level population in LTE. In particular, for a rotational ladder with quantum number J,

$$Q = \sum_J (2J + 1) e^{hBJ(J+1)/kT_{ex}} \simeq \frac{kT_{ex}}{hB}, \tag{7.16}$$

where the last step approximates the discrete sum with an integral.

A example CO $J = 2-1$ spectrum from a dense star-forming region in the Rosette molecular cloud is shown in Figure 7.5. For this transition, $\nu = 230.538$ GHz and $A_{21} = 6.9 \times 10^{-7}$ s^{-1}. Assuming low optical depth, the measured spectrally integrated intensity of 79.5 K km s^{-1} corresponds to a column density $N_{J=2} = 1.2 \times 10^{20}$ m^{-2}. For an excitation temperature of 27 K, we find $Q = 9.8$ and can convert to a total column density, $N_{tot} = 4.3 \times 10^{20}$ m^{-2}.

Fig. 7.5. An example millimeter wavelength molecular spectrum, plotted as brightness temperature against velocity. The solid line shows the CO $J = 2-1$ rotational line and the dashed line shows the same transition of the ^{13}CO isotopologue. The observations were made at the IRAM 30 m radio telescope on Pico Veleta in Spain.

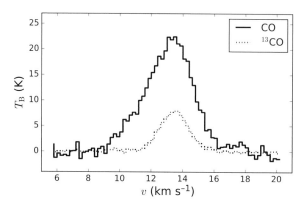

In fact, low rotational transitions of CO in these molecular-rich regions are generally optically thick so these calculations of the column density are a lower limit to the true value. To make a more accurate measure requires observations of an isotopologue. These have the same transitions at nearby frequencies and have similar collisional and radiative excitation rates. Their optical depth, however, scales as the abundance ratio (since the absorption cross-sections are very similar) and is therefore much lower than that of the primary molecule. The dashed line shows the spectrum of the ^{13}CO isotopologue. Despite its relatively low abundance, 70 times less than that of CO, its peak brightness temperature is only about five times lower, which implies that the peak CO line optical depth is very high. Using subscripts 12 and 13 to indicate the primary and isotopologue respectively, the ratio of brightness temperatures relates to the abundance ratio, R, as

$$\frac{T_{B,12}}{T_{B,13}} \simeq \frac{\beta_{12}\tau_{12}}{\beta_{13}\tau_{13}} = R\beta_{12}, \tag{7.17}$$

where we have assumed that the isotopologue emission is optically thin ($\beta_{13} = 1$). This then implies that the escape probability $\beta_{12} \simeq 5/70$, which implies a peak CO optical depth $\tau_{12} \simeq 1/\beta_{12} = 14$.

We can also directly calculate the column density of the ^{13}CO line at a frequency $\nu = 220.399$ GHz and spontaneous emission rate $A_{21} = 6.0 \times 10^{-7}$ s^{-1}. The measured integrated intensity of 18.0 K km s^{-1} implies $N_{J=2} = 2.8 \times 10^{19}$ m^{-2} and, assuming the same excitation temperature, $N_{tot} = 1.0 \times 10^{20}$ m^{-2}. The implied CO column density would be 70 times these values, which is indeed $\sim$14 times higher than the optically thin derived CO column densities above.

To convert to a total molecular column density requires dividing by the CO abundance, [CO]/[H$_2$] $= 10^{-4}$, yielding $N_{H_2} = 7.2 \times 10^{25}$ m^{-2}.

The abundance of different molecules depends on the chemistry of the environment and is discussed later in the chapter.

As with atomic hydrogen regions in Chapter 5, the total mass can then be calculated, $M = N_{H_2} \mu m_{H_2} \Omega d^2$, where Ω is the solid angle over which the column density is calculated, d is the distance to the cloud, and $\mu = 1.35$ accounts for the mass not in hydrogen (primarily helium).

7.6 Temperature

The calculation of the column density in any given transition does not require knowledge of the other level populations but the conversion to a total column density does. Under the assumption of a Boltzmann distribution, the conversion is a simple scaling through the partition function. The excitation temperature can be determined from observations of an optically thick line since then $I_\nu = B_\nu(T_{ex})$. This motivated the choice of 27 K in the calculations above, which used the peak CO temperature plus the 2.7 K cosmic background radiation temperature.

If collisional processes dominate, the excitation temperature is equal to the kinetic temperature and the levels are said to be thermalized. In dense regions where the gas and dust are thermally coupled, the temperature can be estimated from the dust SED (see Chapter 4).

Radiative processes also affect the level population and change the excitation temperature. Below the critical density (Equation 6.30), radiative decay decreases the population of the upper levels resulting in a sub-thermal excitation temperature. Conversely, photon absorption increases the population of the upper levels and optically thick lines can have an enhanced excitation temperature.

The excitation temperature can be measured by comparing the line intensities from different transitions. From Equation 7.15,

$$\ln\left(\frac{N_i}{g_i}\right) = \ln\left(\frac{N_{tot}}{Q}\right) - \frac{E_i}{kT_{ex}}, \qquad (7.18)$$

for each level i. From measurements of the brightness temperature, and therefore column density, at two levels, the two unknowns, T_{ex} and N_{tot}, can be determined. With three or more measurements, a linear regression can be performed on a **rotation diagram** that plots $\ln(N_i/g_i)$ versus E_i/k to determine the slope $-1/T_{ex}$ and intercept $\ln(N_{tot}/Q)$. This is schematically illustrated in Figure 7.6.

The level populations are not always simply described by a single excitation temperature. In such a case, a rotation diagram would not follow a straight line. This may be due to optical depth or gradients in kinetic temperature and density. These effects can each be incorporated into non-LTE models. Such models can be computationally intensive

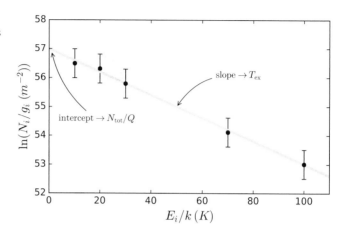

Fig. 7.6. Illustration of a rotation diagram. The points and vertical bars represent observed values and measurement errors. The gray line is a linear fit to the data with a slope and intercept that directly relate to the excitation temperature and total column density of the molecule.

because the absorption of radiation affects the level populations which then affects the emission in different transitions. This inter-dependence is generally solved through iteration. Publicly available codes exist to perform these calculations and can be used for detailed modeling of line profiles in more complex geometries and kinematics than the simple isothermal slab model used above.

7.7 Heating and Cooling

Away from young stars, either in HII regions or embedded protostars, rotational diagrams and other analyses show that molecular gas is generally very cold with temperatures as low as $\sim 10\,\mathrm{K}$ in the densest regions. The reason is the high column densities of gas have high dust extinctions that preclude ultraviolet and optical light from heating the gas.

Dense molecular gas is instead heated by cosmic rays. These relativistic charged particles can penetrate high column densities and their passage induces H_2 ionization. The resulting electron can excite or dissociate other molecules which then exchange energy with the rest of the gas. Cosmic rays are relatively rare and the heating rate is therefore low. The principal cooling agent is CO due to the ease of excitation of rotational modes and its high abundance. There are many details regarding the energy transfer from cosmic rays to kinetic energy of the gas and then to radiation, and many species contribute at different densities and temperatures. However, this basic reasoning shows that the heating will raise the temperature to the observed $\sim 10\,\mathrm{K}$ whereupon the low-lying rotational states of CO and other molecules are excited at a few $10^{-3}\,\mathrm{eV}$ (Figure 7.2).

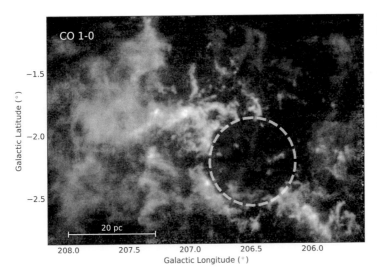

Fig. 7.7. Spectrally integrated CO $J = 1 - 0$ emission toward the Rosette molecular cloud. The dashed circle indicates the approximate location and extent of the HII region discussed in Chapter 6. The CO data are from observations by the (now decommissioned) Five College Radio Astronomy Observatory 14 m radio telescope in Massachusetts.

7.8 Giant Molecular Clouds

Although some molecules had been detected in optical absorption and centimeter emission lines in the decades before, it was the advent of millimeter wavelength astronomy in the 1970s that revealed the extent of the molecular ISM. The 2.7 mm $J = 1-0$ line was first detected toward the Orion nebula (Figure 4.5) and immediately led to the discovery of large and massive molecular complexes. A CO map of one such region, the Rosette molecular cloud, is shown in Figure 7.7.

There is considerable detail in this map but simple, order-of-magnitude estimates are instructive. First, from a point by point integration of spectra at each position in the map to determine column densities and then summing over area, we determine a total mass, $M \simeq 10^5 \, M_\odot$. From the CO spectra in Figure 7.5, we estimate a velocity dispersion along the line of sight of $\sigma \simeq 3 \, \text{km s}^{-1}$. The total kinetic energy, assuming isotropic motions, is therefore $T = \frac{3}{2} M \sigma^2 \sim 3 \times 10^{42} \, \text{J}$. This is very similar to a "spherical cow" estimate of the gravitational energy, $W = GM^2/R \sim 3 \times 10^{42} \, \text{J}$, where we estimate $R = 25 \, \text{pc}$ and assume a uniform density. The balance between kinetic and potential energy suggests that the cloud is a coherent structure with internal motions that are at least partially induced by the gravitational potential.

Observations of polarized dust emission show that dust grains are aligned by a magnetic field (recall Figure 4.10). The Planck satellite made all-sky maps in all Stokes parameters at far-infrared and millimeter wavelengths. Its principal goal to map fluctuations in the CMB required careful removal of all other emission components and resulted in beautiful all-sky maps of Galactic dust emission. Figure 7.8

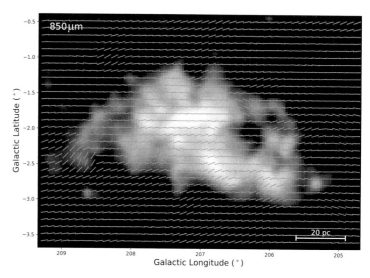

Fig. 7.8. Polarization of dust emission at 850 μm in the Rosette molecular cloud as measured by the Planck satellite. The background image shows the total intensity, where white shows the regions of strongest emission. The lines show the polarization angle rotated by 90° to follow the direction of the magnetic field as projected on the plane of the sky.

shows that the magnetic field is disturbed around the Rosette molecular cloud and correlated with the HII region and more diffuse outer parts. The alignment suggests that the magnetic field is an important factor for understanding molecular cloud properties. The field strength can be measured by comparing the dispersion in polarization angle with linewidth (the Chandrasekhar–Fermi method) or through the **Zeeman effect** which relates fine structure splitting of particular spectral lines, most notably from hydrogen and CN, to the strength of the magnetic field. Field strengths are estimated to be of order a nanotesla. We discuss dynamical considerations in more detail in Chapter 8 and only note here that the implied energy in the magnetic field, $\mathcal{M} \sim \left(\frac{B^2}{2\mu_0}\right)\left(\frac{4}{3}\pi R^3\right) \sim 10^{42}$ J, is comparable to T and W.

The average H_2 number density can be estimated from the column density and size, assuming that the cloud extends as far along the line of sight as its projected size on the sky, $n_{H_2} = 3N_{H_2}/4R \sim 10^8 \, \mathrm{m}^{-3}$. This is similar to the cold neutral atomic medium but the dispersion about the average is much greater. Figure 7.7 shows that molecular clouds have substantial internal structure from small scales to large which can be variously described as clumps or turbulent concentrations. The density variations in the cloud range over more than an order of magnitude above and below this average value.

The densest regions in the cloud, where $n_{H_2} > 10^{10} \, \mathrm{m}^{-3}$, emit strongly in lines from molecules with high critical densities, such as HCN and HCO^+. In these clumps, gravity can overcome thermal pressure and other means of support, which leads to collapse and the formation of stars. The large amounts of dust obscure optical light but infrared

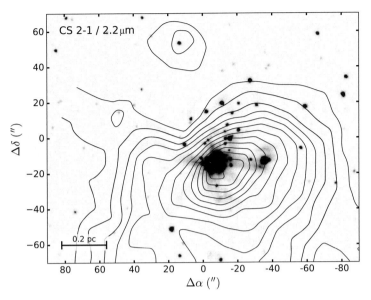

Fig. 7.9. Star formation in a dense clump in the Rosette molecular cloud. The background image is near-infrared K-band emission from a young, embedded cluster of protostars on an inverted scale. The contours show velocity-integrated CS $J = 2-1$ emission from dense gas. A separate core forming a single star is also visible. The infrared image was taken from the UKIRT Infrared Deep Sky Survey (UKIDSS) public release and the CS observations were made at the IRAM 30 m radio telescope on Pico Veleta in Spain.

imaging shows clusters of protostars in dense clumps (Figure 7.9). The prominent Rosette nebula HII region (Figure 6.1), powered by short-lived O stars, also illustrates the close association with young stars.

Away from the dense, star-forming regions, the (optically thick) CO brightness temperatures indicate low kinetic temperatures, $T \sim 10\,\mathrm{K}$. Such low values testify to the wide range of molecular levels and their collective ability to radiate away collisional energy. The mean thermal speed (from the Maxwell–Boltzmann distribution) is $(2kT/m_{CO})^{1/2} \sim 0.08\,\mathrm{km\,s^{-1}}$. Because the observed linewidth is much higher than this, we infer that the cloud is not in thermal pressure equilibrium but is a dynamic entity dominated by turbulent motions. Understanding the source of these motions and their role in cloud structure and star formation are long-standing questions in the field.

The Rosette is an example of a **giant molecular cloud** (GMC), a class of objects with sizes $2R \sim 10-100\,\mathrm{pc}$ and masses ranging from $\sim 10^4\,M_\odot$ to above $10^6\,M_\odot$. There are also numerous lower mass counterparts, termed dark and diffuse clouds, but the $\sim 10^9\,M_\odot$ of molecular gas in the Galaxy is mostly found in the giant objects. The more massive a cloud, the larger it is, with an approximate scaling $M \sim R^2$. This implies a roughly constant average column density that corresponds to a visual extinction of several magnitudes. The velocity dispersion increases with cloud size with an empirically determined scaling, $\sigma \sim R^{0.5}$. These size–mass and size–linewidth relations imply that the overall gravitational potential and kinetic energy both scale with cloud volume, R^3, and, as we found for the Rosette, are similar

Fig. 7.10. Schematic of the size–linewidth relation for the internal structure in clouds (clumps and cores) and for clouds themselves. The velocity dispersion increases approximately as the square root of the size (gray line) across a wide range of scales and smoothly from clouds as a whole to the structures within them.

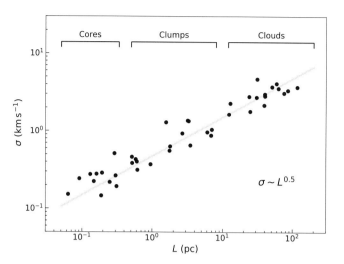

in magnitude. Thus GMCs share similar fundamental properties in lying just above a column density threshold to block energetic radiation that would dissociate molecules and being in approximate dynamical balance.

The power law relation between size and velocity dispersion extends across a wide range of scales from giant to small clouds and to the structures within them, schematically illustrated in Figure 7.10. The boundary between collections of clouds, individual clouds, and clumpy structures within clouds is, by nature, somewhat indistinct and the continuity of the relation suggests a common underlying, probably turbulent, theme whose origin remains poorly understood. The smallest objects, molecular cores, have radii of just a few tenths of a parsec and linewidths of $\sim 0.1 \, \text{km s}^{-1}$, suggesting that they may be supported by thermal pressure. They have high densities, $n_{H_2} \gtrsim 10^{10} \, \text{m}^{-3}$, and are the sites of individual star formation which we discuss in Chapter 9.

The GMCs are the largest, coldest, and densest objects in the ISM. They tend to lie close to the Galactic plane and are associated with the dark patches of nebulosity seen in optical all-sky maps (see Figure 1.1). As with the Rosette cloud and nebula, GMCs are often found to harbor or border HII regions. The energetic radiation from these O stars eventually destroys or disperses the clouds. It is difficult to decipher the azimuthal Galactic structure of the ISM from our vantage point in the plane of the Galaxy, but studies of nearby face-on galaxies show that the HII regions and GMCs lie in spiral arms (see Chapter 11). The compression associated with the passage of the ISM through an arm induces molecular cloud and then star formation. The paucity of molecular gas between arms shows that GMCs have moderately short lifetimes, $\lesssim 30$ Myr. The structure of the ISM on Galactic scales is described in more detail in Chapter 10.

7.9 Photon Dominated Regions

The edge of a molecular region, where molecules turn to atoms, is termed a **photodissociation region** or interchangeably a **photon dominated region**, both with the same PDR abbreviation. Far-ultraviolet radiation (FUV), defined as wavelengths shorter than 200 nm to the hydrogen ionization limit 91.2 nm (energies 6.2–13.6 eV), can dissociate molecular hydrogen, ionize carbon, and broadly affect the physical properties and chemical composition of the gas.

The structure of a PDR is illustrated in Figure 7.11. The FUV radiation field enters the region from the left hand side and is attenuated by dust in the atomic region. Once it drops below a certain level, the H_2 formation rate exceeds its photodissociation rate and the gas becomes predominantly molecular. The H_2 molecules further shield the interior and rarer molecules, first CO and then others, form deeper in the cloud. The spectral lines from the various species regulate the temperature and provide distinct signatures of each region of the PDR.

Calculating the structure and observable properties of PDRs requires detailed models but it all starts with H_2. As a stable molecule, its formation is an exothermic reaction with a binding energy of 4.5 eV. We discuss its formation on dust grain surfaces in the following section and focus here on its excitation and dissociation. The potential energy of H_2 as a function of the distance between its nuclei is illustrated in Figure 7.12. The lower, ground state asymptotes to zero at large separations and has a minimum, equal to negative the binding energy, at

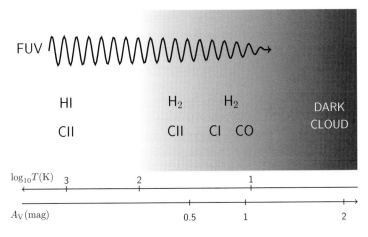

Fig. 7.11. The structure of a PDR. The FUV radiation enters from the left hand side into a neutral atomic cloud and is attenuated by dust to the point where molecular hydrogen begins to form in sufficient numbers to self-shield. Deeper in, carbon becomes neutral and then reacts with oxygen to form CO. Very little radiation penetrates further and more molecules form in the cold, dark interior.

Fig. 7.12. Schematic of the potential energy of an H_2 molecule as a function of its internuclear separation. The lower thick curve represents the ground state and the upper an excited (Lyman or Werner) electronic state. Horizontal lines represent vibrational energy levels and vertical arrows show some possible transitions. Dissociation occurs on the rare occasions when an excited molecule decays into a vibrational continuum of the ground electronic state.

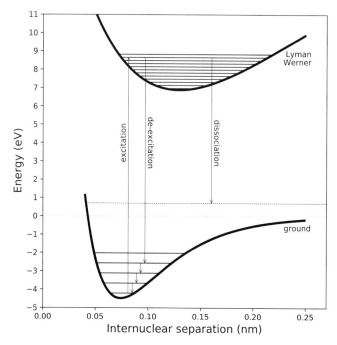

a separation of 0.074 nm. The nuclei repel each other as they are pushed closer leading to the steep increase in the energy at small separations. The nuclei vibrate back and forth at constant energy, shown as horizontal lines in this diagram with discrete steps described in Equation 7.3 (and with unshown finer scale rotational levels as in Equation 7.4). The upper thick line represents a higher energy state where the electrons are in an excited orbital. The potential has the same general shape, for the same reasons, though it is broader and with a minimum at slightly wider nuclear separation. In the ISM, where atomic hydrogen effectively absorbs all photons with energies $> 13.6\,\text{eV}$, the two electronic states that can be excited are the Lyman and Werner bands with minimum energies of 11.4 and 12.4 eV respectively.

A hydrogen molecule in an interstellar cloud will be in the ground electronic and vibrational state. Dissociation proceeds through excitation to one of the Lyman or Werner bands and spontaneous decay into a vibrational continuum at positive potential energy. However, the more likely transitions are into a stable vibrational state in the ground electronic level, which then subsequently cascades to the lowest state. Only about 10–15% of H_2 excitations (dependent on the spectrum of the incident radiation field) break the molecule apart.

The low efficiency of photodissociation means that the formation of a moderate amount of H_2 blocks deeper penetration of FUV radiation.

This is known as **self-shielding** and results in a rapid cloud transition from atomic to molecular at a visual extinction, $A_V \sim 0.3$ mag. The line absorption is limited to a discrete set of wavelengths but, as the H_2 column density increases, the equivalent width increases and the windows in which the FUV passes through narrow. Eventually the combination of dust and H_2 shielding attenuates the radiation field sufficiently for the ionized carbon to become neutral and then to react with oxygen to form the next most abundant molecule, CO, at $A_V \sim 1$ mag. For a cloud that is illuminated on both sides, these extinction thresholds should be doubled.

The PDRs can be very bright due to the combination of incident radiation field with moderate densities and temperatures at the atomic–molecular interface. This produces strong thermal dust continuum and atomic emission lines in the far-infrared, as well as numerous ro-vibrational lines of excited H_2 and high J rotational lines of CO ranging through the mid-infrared to the millimeter wavelength regimes. The line strengths depend mainly on the radiation field and gas density, and are therefore diagnostics of these PDR properties.

Hydrogen molecules can also be dissociated through collisions which can occur in shocks (Chapter 8) or through thermal motions at temperatures $\gtrsim 4000$ K. The energy that goes into breaking the molecular bond acts as a strong coolant and prevents the molecular gas from rising above this value. At lower temperatures, the primary coolants are fine structure lines of CII at 158 µm with a critical density of 3×10^9 m^{-3}, and OI at 63 µm with a critical density of 5×10^{11} m^{-3}. Deeper into the PDR, CI at 609 µm and CO rotational lines take over. As the gas cools, it becomes denser and transitions from a PDR to a dark cloud with a rich molecular chemistry.

Although most of the *volume* of the ISM is occupied by low-density, ionized hydrogen, the FUV radiation field is dilute in much of the Galaxy due to the large distances between stars. Consequently, much of the *mass* of the ISM is neutral with sufficient dust extinction for these PDR considerations to apply. Understanding how such regions are affected by different radiation fields and amounts of dust is also essential for understanding other galaxies.

7.10 Astrochemistry

The detection of molecules in the ISM naturally leads to the question of their origins and their reactions with each other. Astrochemistry is quite different from its terrestrial counterpart on account of the low densities and long timescales.

A stable molecule, XY, requires energy to be dissociated and must therefore be in a lower energy state than its individual components. Therefore, for X and Y to bind together requires that the collisional and formation energy be removed from a temporarily excited state XY^*. The lifetime of this excited state is very short, of order the vibrational timescale, $t^* \sim 1/\omega_{vib} \sim 10^{-13}$ s. In dense environments such as our or other planetary atmospheres, the energy can be carried away by a third particle, but such almost instantaneous three-body encounters are extremely rare in the ISM. Consider an atomic cloud with $n_H \sim 10^8$ m^{-3}, $T \sim 100$ K (i.e., the CNM). Collisions with the dominant species, H, occur on a timescale $t_{coll} \sim 10^9$ s so the likelihood of a third-body encounter with the excited state XY^* is a vanishingly small $t^*/t_{coll} \sim 10^{-22}$.

Instead, most ISM chemical pathways in the gas phase proceed through **ion–molecule reactions**,

$$X^+ + YZ \rightarrow X + YZ^+. \tag{7.19}$$

The cross-section and timescale for interaction are increased in this case because the ion induces a dipole moment in the molecule. As carbon has a lower ionization potential than hydrogen, it is abundant in diffuse clouds and is an important chemical reagent. Cosmic rays penetrate well beyond starlight into highly extincted dark clouds and their high energies leave a trail of ionized species along their track.

However, ion–molecule reactions require a molecule to exist so we must look elsewhere to understand how astrochemistry starts. This leads us to consider reactions on the surfaces of dust grains. Dust provide a stable location for atoms to reside, weakly bound by van der Waals forces, and absorbs the energy produced by any reaction. These **grain surface reactions** can be written as a two- or three-stage process,

$$X + S \rightleftharpoons XS$$
$$Y + XS \rightarrow XY + S, \tag{7.20}$$

$$X + S \rightleftharpoons XS$$
$$Y + S \rightleftharpoons YS \tag{7.21}$$
$$XS + YS \rightarrow XY + S.$$

In the first case, species X bonds tightly to the grain surface S (or is too massive to move) and species Y reacts directly with this bond from the gas phase. In the second case, both species, X and Y, bond more loosely to the surface and diffuse or "hop" around until they encounter each other and react. Either way, the grain is left unchanged after the reaction and acts only as a catalyst.

The three-stage grain surface process is the dominant pathway for the formation of H_2. To estimate how fast this occurs, we start with the collisional timescale of hydrogen atoms onto grain surfaces,

$$t_{H-d} = \frac{1}{n_d \sigma_d v_H} \simeq \frac{N_H}{A_V} \frac{1}{n_H} \left(\frac{m_H}{kT}\right)^{1/2}$$
$$= 8 \left(\frac{n_H}{10^8 \text{m}^{-3}}\right)^{-1} \left(\frac{T}{100 \text{ K}}\right)^{-1/2} \text{Myr}, \qquad (7.22)$$

where we have multiplied by a length scale to convert the dust terms to an optical depth, used the empirical relation between visual extinction and hydrogen column density in Equation 5.24, and normalized to the typical properties of the CNM. The H_2 formation time is then

$$t_{H_2,\text{form}} = \frac{2t_{H-d}}{\epsilon_{H_2} P_S}, \qquad (7.23)$$

where P_S is the probability that a hydrogen atom sticks to the grain surface and ϵ_{H_2} is the reaction efficiency between pairs of hydrogen atoms. At the cold temperatures of dust grains in dark clouds, $P_S \simeq 1$, and the reaction also occurs with almost 100% efficiency, $\epsilon_{H_2} \simeq 1$. Thus the H_2 formation timescale in the CNM is typically about 16 Myr, a relatively slow process due principally to the low number density of dust grains.

This formation timescale is comparable to the photodissociation time in diffuse gas but they vary in opposite ways as the density increases. The formation time decreases due to the faster collisional rate whereas the dissociation time increases sharply due to dust extinction of ultraviolet radiation. By $A_V \simeq 0.3$, the H_2 formation rate exceeds the dissociation rate and the cloud begins to become molecular (Figure 7.11). Self-shielding acts to further reduce the photodissociation rate.

Once H_2 forms, astrochemistry begins in earnest. The first step is ionization by cosmic rays (CR) and the formation of protonated molecular hydrogen,

$$H_2 + CR \rightarrow H_2^+ + e^- + CR$$
$$H_2^+ + H_2 \rightarrow H_3^+ + H \qquad (7.24)$$

This is an exothermic, fast reaction and H_3^+ is a stable molecule that readily donates a proton to other molecules to jump-start a chain of gas-phase ion–molecule reactions that ends with CO, H_2O, and numerous other species. These two principal processes that dominate the chemistry as an atomic cloud turns molecular are schematically represented on the left hand side of Figure 7.13.

To model the full chemistry of a molecule requires balancing the full set of reactions involving its varied formation and destruction pathways. This generally involves other molecules and leads to an extended chemical network that includes the inter-related formation and destruction

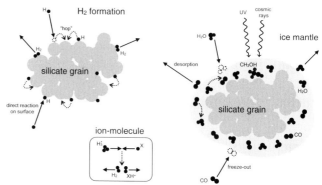

Fig. 7.13. Illustration of chemical reactions in the ISM. The left hand side shows gas phase and grain surface reactions that occur as an atomic cloud turns molecular. The right hand side applies to the conditions inside a cold, dense molecular core. The grain is surrounded by an ice mantle of water and carbon monoxide in the cold ISM, $T \ll 100\,\mathrm{K}$. Ultraviolet radiation and cosmic rays promote reactions within this mantle that lead to larger, more complex molecules that may then desorb or otherwise release into the gas. For a sense of scale (not shown in these cartoons), typical silicate grain sizes are $\sim 0.1\ \mu$m and the ice mantle may be $\sim 10^2$ monolayers thick.

of other species. Since the abundance of a given molecule depends directly on the abundance of the individual atoms that it is composed of, the steady-state equilibrium solution requires solving a series of linear equations through matrix inversion. To model the time evolution of all species requires solving a coupled set of ordinary differential equations.

Chemical models can be tested through measurements of column densities and the determination of relative abundances between different species. As the H_2 is not directly observable in the cold gas, proxies such as dust or a CO isotopologue are often used to convert to an absolute abundance, $X = n_X/n_{H_2}$. In relative terms, CO is very abundant, $X \sim 10^{-4}$, and molecular cloud observations routinely map the distribution of CS with $X \sim 3 \times 10^{-9}$, and even rarer species such as N_2H^+ with $X \sim 10^{-10}$. The faintest lines in Figure 7.4 are glycolaldehydes (composed of H, C, and O) with abundances $X \sim 10^{-11}$.

In cold, dense cores of molecular clouds, volatile molecules such as H_2O and CO freeze out onto grains to produce icy mantles which themselves open up new chemical pathways (illustrated on the right hand side of Figure 7.13). The ices can be seen through their vibrational modes as absorption lines against embedded or background infrared sources (Figure 7.14). Cosmic rays can directly impart energy onto grains that power additional surface reactions. The recombination of the ions that they produce creates a diffuse FUV field deep in the cloud, which similarly helps hydrogenate CO to produce relatively long chain

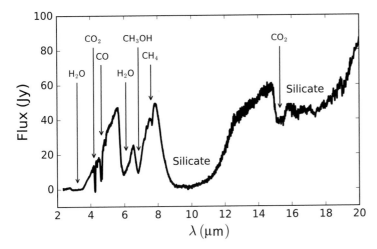

Fig. 7.14. Mid-infrared absorption spectrum toward the massive star-forming region W33A, from the Infrared Space Observatory (ISO) archive. The most prominent, broad absorption features are due to SiO stretching and bending modes in silicate minerals in the grain interior. The other features arise from ices on the grain surface that are created by molecules freezing out of the gas phase and undergoing subsequent reactions.

organic compounds, such as methanol (CH_3OH), that are unlikely to form in the gas phase alone. Comets ultimately originate from icy grains in circumstellar disks so understanding the properties of these ices and their heritage from the ISM is of great importance to planet formation and astrobiology.

7.11 Dust as Big Molecules

The largest molecule detected in the ISM to date is ionized Buckminsterfullerene (C_{60}^+). Carbon is one of the more common elements in the Universe and is able to bond with itself to form long chains, known as catenation. Attached to hydrogen, this forms a class of compounds known as **polycyclic aromatic hydrocarbons** (PAHs) or, more colloquially in the terrestrial context, "soot". There are so many bending and stretching modes possible in PAHs that any single line is generally very weak but collectively they can produce a near continuum of diffuse emission at $\lambda \sim 20-200\ \mu m$, known as **infrared cirrus**.

The size of such large molecules can be several nanometers and is comparable to that of the smallest dust grains described in Chapter 4. This brings us full circle in our description of the different components of the ISM. We now move on to dynamics and the role of the ISM in galaxy evolution and star formation.

Notes

A comprehensive online reference for molecular spectroscopy is spec.jpl.nasa.gov. The Leiden atomic and molecular database also provides emission rates and collisional cross-sections for commonly observed species in the ISM and is at home.strw.leidenuniv.nl/~moldata. This website also links to an online program, RADEX, for calculating the radiative transfer in non-LTE conditions. The rich spectrum at millimeter wavelengths in Figure 7.4 is part of the PILS survey, which can be explored further at youngstars.nbi.dk/PILS. It was provided to the author by Jes Jorgensen. The Planck map of polarized dust emission in the Rosette molecular cloud was published by Planck Collaboration et al. (2016). The data were provided to the author by Marta Alves. For more detail on the topics of PDRs and astrochemistry that are only lightly covered here, see the textbook by Tielens (2010) and the review by Herbst and van Dishoeck (2009).

Questions

1a. Consider a diatomic molecule with rotational constant B. Give an expression for J_{max}, the level that has the highest population as a function of temperature assuming LTE conditions. Plot the relative population levels, N_J/N_{tot}, for the CO molecule ($B = 57.6$ GHz) for $T = 10, 30, 100$ K.

1b. Suppose you observe the $J = 3 - 2$ and $J = 1 - 0$ CO lines of a molecular cloud. If the line strengths are the same, what can you conclude about the cloud temperature?

2. Estimate the average thermal pressure in a GMC. How does it compare to the CNM/WNM? Explain any discrepancy.

3. The integrated intensity of the CS $J = 2 - 1$ emission in the large clump in Figure 7.9 is 10 K km s^{-1}. The embedded stellar cluster warms the gas to 50 K. Assuming that the abundance of CS relative to H_2 is 10^{-8}, estimate the total mass of gas.

4. What observational probes exist for a PDR between $A_V \sim 2-4$ mag where the only molecule is H_2?

Chapter 8
Dynamics

The ISM is constantly in motion. The turbulence in molecular clouds is balanced by their self-gravity. Thermal pressure supports the diffuse atomic and ionized gas on large scales but massive stars heat the gas and cause it to expand. The impulse of combined supernova explosions can affect gas on kiloparsec scales. To reveal these dynamics of the ISM, we start with the equations of fluid mechanics.

8.1 Fluid Mechanics

As noted in the introductory Chapter 1, although the collisional rate is low compared to many spontaneous emission rates in the ISM, the mean free path,

$$l = \frac{1}{n\sigma} \simeq 5 \times 10^{-3} \left(\frac{n_{\rm H}}{10^6\,{\rm m}^{-3}}\right)^{-1} {\rm pc}, \tag{8.1}$$

is much smaller than the clouds and HII regions under study here. Consequently we can define macroscopic quantities such as density ρ, velocity $\mathbf{v}$, and pressure P, and use fluid equations to relate them to each other.

If we consider an individual volume fixed in space, its properties can change due to the flow of material across its surface. To simplify the visualization, Figure 8.1 shows a one-dimensional flow of a compressible fluid through a small region of length Δx with a fixed area A in the other two dimensions.

The mass within the volume, $M = \rho A \Delta x$, changes over time interval Δt due to the difference between the material flowing in and out,

$$\Delta M = \rho A v \Delta t - (\rho + \Delta\rho)A(v + \Delta v)\Delta t. \tag{8.2}$$

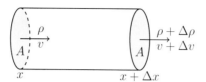

Fig. 8.1. The flow of a fluid through a fixed volume in space. Material enters from the left side with a density ρ and velocity v, and leaves from the right side with a change in these values of $\Delta\rho, \Delta v$, respectively.

In the limit of small quantities, this leads to the differential form

$$d\rho dx = -d(\rho v)dt, \tag{8.3}$$

which generalizes in three dimensions with vector notation to the equation of mass conservation,

$$\frac{\partial \rho}{\partial t} + \nabla.(\rho\mathbf{v}) = 0. \tag{8.4}$$

In a similar way, the momentum of the material in the volume changes due to the flow through it and the forces acting on it. This leads to the equation of momentum conservation,

$$\frac{\partial \mathbf{v}}{\partial t} + (\mathbf{v}.\nabla)\mathbf{v} = -\frac{\nabla P}{\rho} - \nabla\phi. \tag{8.5}$$

Here ϕ is the gravitational potential of the fluid, which depends on the density via Poisson's equation,

$$\nabla^2\phi = 4\pi G\rho. \tag{8.6}$$

A fourth equation describes the evolution of the energy of the system and is expressed through the **equation of state**,

$$P = nkT, \tag{8.7}$$

where P is the pressure of the fluid, T its temperature, and $n = \rho/m$ is the number density for particles of mass m. Depending on how quickly the gas radiatively cools relative to dynamical changes in the macroscopic quantities, there are two limiting cases, **isothermal** and **adiabatic**, where P depends only on ρ and the flow is said to be barotropic.

In the isothermal case, the cooling time is short and any energy input to the system is quickly radiated away. The gas has a constant temperature so pressure and density changes are proportional to each other,

$$P_{\text{isothermal}} \propto \rho. \tag{8.8}$$

The adiabatic case is the opposite extreme where cooling is relatively slow compared to dynamical timescales in the system and changes in pressure are converted to thermal energy. If the gas particles have n_f

degrees of freedom, the total internal energy $U = N \times n_f kT/2 = n_f PV/2$, where $N = nV$ is the total number of particles in volume V. Changes in the pressure and volume are related to each other via the first law of thermodynamics which, for an adiabatic change, takes the form

$$dU = -PdV = n_f(PdV + VdP)/2$$
$$\Rightarrow dP/P = -\gamma \, dV/V, \quad \text{where } \gamma = 1 + 2/n_f$$
$$\Rightarrow P_{\text{adiabatic}} \propto V^{-\gamma} \propto \rho^\gamma. \quad (8.9)$$

For monoatomic gas, $n_f = 3$ and $\gamma = 5/3$. Diatomic molecules have an extra two degrees of freedom, $n_f = 5$ so $\gamma = 7/5$. The adiabatic case behaves similarly to the isothermal case for particles with many degrees of freedom, $n_f \gg 1 \rightarrow \gamma \simeq 1$, as there are many ways to store changes in the energy.

The simplicity of these equations belies the complexity of fluid motions. Nevertheless, in limited circumstances, analytic solutions exist and provide important physical insights.

8.2 The Wave Equation

What happens when a fluid at rest undergoes small perturbations, such as dropping a pebble into a pond? We first consider the case with negligible gravity and, for simplicity, return to the one-dimensional case, though the following is readily generalized using vector calculus.

Labeling the static solution and perturbations with subscripts 0 and 1 respectively, the variables are

$$v = v_1, \quad \rho = \rho_0 + \rho_1, \quad P = P_0 + P_1. \quad (8.10)$$

This approach makes the non-linear term in the momentum conservation equation a second-order term which can be neglected. The perturbations are then related in a near-symmetric way,

$$\frac{\partial \rho_1}{\partial t} + \rho_0 \frac{\partial v_1}{\partial x} = 0, \quad (8.11)$$

$$\frac{\partial v_1}{\partial t} + \frac{1}{\rho_0} \frac{\partial P_1}{\partial x} = 0. \quad (8.12)$$

The small pressure change, P_1, induces a density change, ρ_1 (or vice versa). If the perturbations behave adiabatically, the equation of state relates the two via

$$\frac{P_0 + P_1}{P_0} = \left(\frac{\rho_0 + \rho_1}{\rho_0} \right)^\gamma$$
$$\Rightarrow P_1 \simeq \left(\frac{\gamma P_0}{\rho_0} \right) \rho_1 \equiv a^2 \rho_1, \quad (8.13)$$

where $a^2 = \gamma P_0/\rho_0 = \gamma kT/m$ is the adiabatic **sound speed**. Now differentiating Equation 8.11 with respect to t and Equation 8.12 with respect to x and subtracting gives

$$\frac{\partial^2 \rho_1}{\partial t^2} - a^2 \frac{\partial^2 \rho_1}{\partial x^2} = 0. \tag{8.14}$$

This is known as the wave equation and has the general solution $\rho_1 = f(x \pm at)$, where f is any function (try it and see for yourself). What this means is that any small perturbation to the fluid equations at (x_1, t_1) looks the same at (x_2, t_2), where $x_1 \pm at_1 = x_2 \pm at_2$. That is, the pressure and density perturbation travels in space at the sound speed $x/t = \pm a$.

The same form applies for the isothermal case and the terminology distinguishes the sound speed from the adiabatic case by labeling $c = (kT/m)^{1/2}$.

8.3 The Jeans Criterion

What happens when a self-gravitating cloud in the ISM undergoes small perturbations, such as being nudged by an HII region? Named after its discoverer, the Jeans criterion extends the perturbation approach to a static solution by relating the change in gravitational potential to density via $\nabla^2 \phi_1 = 4\pi G \rho_1$. With an isothermal equation of state, $P = \rho c^2$, and staying in vector notation, the end result is the wave equation modified by a gravitational term,

$$\frac{\partial^2 \rho_1}{\partial t^2} = c^2 \nabla^2 \rho_1 + 4\pi G \rho_0 \rho_1. \tag{8.15}$$

To determine the nature of the solution, we assume a wavelike solution,

$$\rho_1 \propto e^{i(\omega t - \mathbf{k} \cdot \mathbf{x})}. \tag{8.16}$$

The use of complex notation here is a compact shortcut that simplifies the algebra and results in the **dispersion relation** between temporal and spatial wavenumbers,

$$\omega^2 = k^2 c^2 - 4\pi G \rho_0 \equiv (k^2 - k_J^2)c^2, \tag{8.17}$$

where the Jeans wavenumber $k_J = (4\pi G \rho_0/c^2)^{1/2}$.

If $k > k_J$ then ω has a real solution and the perturbation is a gravity-modified sound wave as in Equation 8.14. If, however, $k < k_J$ then $\omega^2 < 0$ so ω is a (positive and negative) imaginary number and the perturbation grows exponentially with time. This condition on wave number translates to length scales,

$$L > L_J = \frac{2\pi}{k_J} = \left(\frac{\pi}{G\rho_0}\right)^{1/2} c, \tag{8.18}$$

where L_J is the Jeans length. We will see in the following chapter that the term in brackets above is proportional to the free-fall timescale for gravitational collapse. The above inequality can therefore be restated to compare this timescale to the crossing time of a sound wave, L/c. In other words, for perturbations on scales greater than the Jeans length, the pressure cannot respond fast enough to resist gravitational growth. The Jeans length translates to a Jeans mass,

$$M_J = \frac{4}{3}\pi\rho_0 \left(\frac{L_J}{2}\right)^3 = \frac{1}{6}\left(\frac{\pi^5}{G^3\rho_0}\right)^{1/2} c^3. \tag{8.19}$$

The Jeans criterion states that, under the assumption of a constant density and temperature, an object that exceeds these size or mass limits will undergo gravitational collapse. For the typical densities and temperatures of a molecular core, the Jeans mass is comparable to that of the Sun (Chapter 9). As the mass of a star is the basic determinant of its evolution, lifetime, and end state, this connection to observed properties of the ISM is of great interest.

The Jeans length and mass both scale inversely with the square root of the density. This suggests that a collapsing cloud may break into multiple smaller pieces as it becomes denser, a concept known as Jeans fragmentation. Indeed, stars (and galaxies) tend to form in groups. However, the situation is more complex than this simple reasoning as the initial conditions of a static system no longer apply. Furthermore, the very nature of a stable, self-gravitating, fluid that can be perturbed is physically unrealizable. In terms of the Poisson equation, the only solution to $\phi_0 = $ constant is for $\rho_0 = 0$. Ignoring this inconvenience is known as the "Jeans swindle", but the concept of a critical length and mass scale still hold, and can be derived in particular cases with more appropriate equilibrium conditions (see Chapter 9).

8.4 The Virial Theorem

The fluid equations above were derived by consideration of flows across the surface of a volume fixed in space, a so-called Eulerian description. It is equally valid to track an individual fluid element as it moves through space such that its properties change both with time and along its path. We can transform between the two descriptions through the use of the Lagrangian derivative,

$$\frac{D}{Dt} \equiv \frac{\partial}{\partial t} + \mathbf{v}.\nabla, \tag{8.20}$$

where the second term here accounts for convection in and out of the spatial element.

In this Lagrangian description, the momentum conservation Equation 8.5 takes a simpler form that looks like Newton's second law,

$$\rho \frac{D\mathbf{v}}{Dt} = -\nabla P - \rho \nabla \phi. \tag{8.21}$$

This is the starting point for deriving the virial theorem which relates the kinetic, thermal, and gravitational energies of a fluid (for simplicity, we ignore pressure terms henceforth).

D/Dt is simply an operator that can be applied to scalars or vectors and it has the same behavior as any derivative. Using the identity

$$\frac{D^2 r^2}{Dt^2} = \frac{D}{Dt}\left(\frac{D}{Dt}\mathbf{r}.\mathbf{r}\right) = \frac{D}{Dt}2\mathbf{r}.\mathbf{v} = 2v^2 + 2\mathbf{r}.\frac{D\mathbf{v}}{Dt}, \tag{8.22}$$

we can convert the pressureless form of Equation 8.21 to

$$\frac{1}{2}\rho\frac{D^2 r^2}{Dt^2} - \rho v^2 = -\rho\mathbf{r}.\nabla\phi. \tag{8.23}$$

We now integrate over volume. By keeping the density here, this is equivalent to summing over the masses of each fluid element, $dm = \rho dV$, which makes the physical interpretation clearer,

$$\frac{1}{2}\frac{D^2 I}{Dt^2} - 2T = -\int \mathbf{r}.\nabla\phi \, dm, \tag{8.24}$$

where $I = \int r^2 dm$ is the moment of inertia, and $T = \int \frac{1}{2}v^2 dm$ is the kinetic energy. The right hand term requires a bit more manipulation to understand its nature. First, we recognize that $\nabla\phi dm$ is a gravitational force vector. In the absence of an external gravitational field acting on the system, we can convert this term to an analogous sum over discrete particles,

$$W = -\int \mathbf{r}.\nabla\phi \, dm \rightarrow -\sum_i \sum_{j \neq i} \mathbf{r_i}.\mathbf{F_{ij}}. \tag{8.25}$$

The double sum can be broken up into two parts and the second part re-arranged to count the same pairs of particles in a different order as follows,

$$\sum_i \sum_{j \neq i} \mathbf{r_i}.\mathbf{F_{ij}} = \sum_i \sum_{j < i} \mathbf{r_i}.\mathbf{F_{ij}} + \sum_i \sum_{j > i} \mathbf{r_i}.\mathbf{F_{ij}}$$

$$= \sum_i \sum_{j < i} \mathbf{r_i}.\mathbf{F_{ij}} + \sum_j \sum_{i < j} \mathbf{r_i}.\mathbf{F_{ij}}$$

$$= \sum_i \sum_{j < i} \mathbf{r_i}.\mathbf{F_{ij}} + \sum_i \sum_{j < i} \mathbf{r_j}.\mathbf{F_{ji}}. \tag{8.26}$$

The last step simply swaps the labels i and j, which then has the nice result that the two parts now sum over the same set of indices. Finally we substitute in the expression for gravitational force between two particles,

$$\mathbf{F_{ij}} = \frac{Gm_im_j(\mathbf{r_i} - \mathbf{r_j})}{|\mathbf{r_i} - \mathbf{r_j}|^3}, \tag{8.27}$$

and we see that

$$W = -\sum_i\sum_{j<i}(\mathbf{r_i} - \mathbf{r_j}).\,\mathbf{F_{ij}} = -\sum_i\sum_{j<i}\frac{Gm_im_j}{|\mathbf{r_i} - \mathbf{r_j}|} \tag{8.28}$$

is the gravitational potential energy of the system.

A fluid, or collection of particles that acts as such, is said to be **virialized** when $D^2I/Dt^2 = 0$, i.e., when the moment of inertia is constant or only changes linearly with time. In this case, the gravitational potential energy is (in magnitude) twice the kinetic energy,

$$2T + W = 0. \tag{8.29}$$

This is a simple, powerful, equation. If we consider a uniform spherical cloud with mass M, radius R, and linewidth σ, then $T = \frac{3}{2}M\sigma^2, W = \frac{3}{5}GM^2/R$, and, if the cloud is virialized,

$$M = \frac{5R\sigma^2}{G}. \tag{8.30}$$

Thus we can estimate masses from the most basic observables of size and linewidth. We can then compare with mass estimates from other means, e.g., line strength or dust extinction for GMCs, and test whether the object is virialized or not. The virial theorem has broad applicability beyond the ISM, from Galactic stellar dynamics to clusters of galaxies.

8.5 Magnetic Fields

The ISM is threaded by magnetic fields, $\mathbf{B}$. The flow of charged particles through the field creates an electric current density, $\mathbf{j} = \frac{1}{\mu_0}\nabla\times\mathbf{B}$, where $\mu_0 = 4\pi\times10^{-7}\,\mathrm{m\,kg\,s^{-2}\,A^{-2}}$ is the magnetic constant, and this then induces a Lorentz force term,

$$\mathbf{j}\times\mathbf{B} = \frac{1}{\mu_0}(\mathbf{B}.\nabla)\mathbf{B} - \nabla\left(\frac{B^2}{2\mu_0}\right), \tag{8.31}$$

in the momentum conservation Equation 8.5. The first term here is a tension that acts to keep force lines straight and the second term is a pressure that resists compression of the field lines (Figure 8.2). Note that there is no force along the field lines since the Lorentz force projects to zero, $\mathbf{B}.(\mathbf{j}\times\mathbf{B}) = \mathbf{j}.(\mathbf{B}\times\mathbf{B}) = \mathbf{0}$.

The tension of bent magnetic fields resists compression along the direction of the field lines. The pressure term resists compression in the

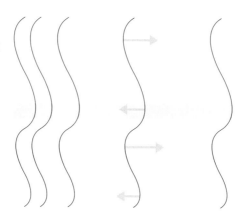

Fig. 8.2. The vertical wavy lines indicate the direction and density of magnetic field lines. The small arrows around a single line represent the tension term which is strongest in regions of greatest gradient and acts to straighten the field lines. The large arrow represents the pressure term which produces a force from strong to weak field-line density.

perpendicular direction. By balancing the magnetic pressure, $P_{mag} = B^2/2\mu_0$, to the kinetic energy density, $\rho v_A^2/2$, we can heuristically define the **Alfvèn speed**,

$$v_A = \frac{B}{(\mu_0\rho)^{1/2}}. \tag{8.32}$$

This is the characteristic speed at which perturbations propagate along magnetic field lines, and is analogous to the sound speed of pressure waves in a fluid. The situation is more complicated than isotropic thermal pressure, however, and the actual speed of magnetic waves depends on their nature (transverse or longitudinal) and the angle of propagation to the field lines.

Observations show that magnetic fields in atomic and molecular clouds are about a nanotesla (10^{-9} kg A^{-1} s^{-2} $= 10\,\mu$G in the cgs notation of much of the literature) and increase with density. Dust polarization maps show ordered fields on large scales in diffuse gas and demonstrate the importance of magnetic fields in ISM structure and dynamics. On the scale of a molecular cloud with $\langle n_{H_2}\rangle \sim 10^8$ m^{-3}, the inferred Alfvèn speed, $v_A \simeq 2$ km s^{-1}, is comparable to the measured linewidth. This also suggests that the turbulent motions in clouds, which are much greater than the thermal sound speed, may be magnetic in nature.

The inclusion of magnetic fields in the fluid equations is the field of **magnetohydrodynamics** and is important not only for the ISM but also a wide variety of astrophysical phenomena from studies of the Sun to jets emanating from the central black holes in active galaxies.

8.6 Shocks

To within 50% (depending on the mean mass of the gas and whether it is molecular, atomic, or ionized), the adiabatic sound speed

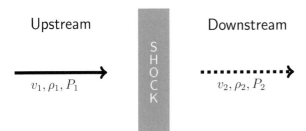

Upstream

Downstream

S
H
O
C
K

v_1, ρ_1, P_1

v_2, ρ_2, P_2

Fig. 8.3. Schematic of a planar shock showing the near-instantaneous change in fluid properties. In the frame of the shock, material moves in at speed v_1 and leaves at speed v_2.

$a = (\gamma P / \rho)^{1/2} \simeq 0.9 \, (T / 100 \, \text{K})^{1/2}$ km s^{-1}. This is relatively small compared with typical ISM motions, even in HII regions with $T \simeq 8000$ K. For such **supersonic** speeds, pressure waves cannot communicate disturbances from one part of the ISM to another and, unless mediated by magnetic Alfvèn waves, this results in abrupt changes in gas properties on small scales. Examples of such **interstellar shocks** include protostellar outflows, HII regions, supernovae, and spiral arms.

In general, the shock is very thin, comparable to the mean free path, and we can simply relate the properties of the gas upstream from the shock with those downstream via conservation of mass and momentum. Consider variables defined in the frame of the shock, as defined in Figure 8.3. The mass flow per unit area per unit time through the shock must be balanced,

$$\rho_1 v_1 = \rho_2 v_2. \tag{8.33}$$

Similarly the momentum flux per unit area per unit time is $\rho_1 v_1^2$ and must be balanced by the difference in the force per unit area (pressure),

$$\rho_1 v_1^2 + P_1 = \rho_2 v_2^2 + P_2. \tag{8.34}$$

These two jump conditions can also be directly derived by integrating the static, one-dimensional fluid equations 8.4 and 8.5, and are often referred to by the combined names of the pioneers in this work, Rankine–Hugoniot. Note that the equations are symmetric but the physics is not due to the large, irreversible change of entropy across the shock.

To solve these equations requires an additional condition that relates the three variables. There are two regimes (and, as usual, a more complex situation in between): an adiabatic case when the cooling timescale is long compared to dynamical changes and little energy is lost to radiation; and an isothermal case when cooling timescales are relatively short and the temperature of the pre- and post-shock gas is the same.

Adiabatic Shocks

Assuming radiative losses are small, then we can balance the energy of the gas with the work done by the shock. The energy density in the upstream flow is the sum of kinetic and internal terms, $E_1 = \frac{1}{2}\rho_1 v_1^2 + U_1$, where

$$U_1 = N_1 \frac{n_f}{2} \frac{kT_1}{V} = \frac{P_1}{\gamma - 1} \tag{8.35}$$

and n_f is the number of degrees of freedom, related to the adiabatic index by $\gamma = 1 + 2/n_f$. The energy per unit area per unit time is then $E_1 v_1$. The work done per unit area per unit time is force×distance/(area×time) = pressure×velocity = $P_1 v_1$. Balancing the sum of these across the shock gives

$$v_1 \left[\frac{1}{2}\rho_1 v_1^2 + \frac{P_1}{\gamma - 1} + P_1 \right] = v_2 \left[\frac{1}{2}\rho_2 v_2^2 + \frac{P_2}{\gamma - 1} + P_2 \right]. \tag{8.36}$$

Note that this assumes γ does not change across the shock, i.e., the physical state of the gas remains the same. It is easy to include this change though it complicates the algebra. We proceed by factoring out the density and using the mass conservation equation to obtain

$$\frac{1}{2}v_1^2 + \frac{\gamma P_1}{(\gamma - 1)\rho_1} = \frac{1}{2}v_2^2 + \frac{\gamma P_2}{(\gamma - 1)\rho_2}. \tag{8.37}$$

We see that the second term of the left hand side is related to the sound speed, $a_1^2 = \gamma P_1/\rho_1$. On the right hand side, we substitute $v_2 = \rho_1 v_1/\rho_2$ and $P_2 = P_1 + \rho_1 v_1^2 - \rho_2 v_2^2$. This leads to a quadratic for the density contrast, ρ_2/ρ_1 in terms of γ and the **Mach number**,

$$M_1 \equiv \frac{v_1}{a_1}. \tag{8.38}$$

One solution of the quadratic is the (trivial) solution $\rho_2/\rho_1 = 1$. The other, more interesting, solution is

$$\frac{\rho_2}{\rho_1} = \frac{(\gamma + 1)M_1^2}{(\gamma + 1) + (\gamma - 1)(M_1^2 - 1)}. \tag{8.39}$$

Note that this asymptotes to a finite value for the case of a "strong shock",

$$\frac{\rho_2}{\rho_1} \to \frac{(\gamma + 1)}{(\gamma - 1)} \quad \text{as} \quad M_1 \to \infty. \tag{8.40}$$

For atomic gas with three degrees of freedom, $\gamma = 5/3$, and the maximum density contrast is 4. The physical reason for this finite value is that without radiative losses the gas heats up and the thermal pressure resists compression.

A little more algebra shows that the pressure jump

$$\frac{P_2}{P_1} = \frac{2\gamma}{\gamma + 1} M_1^2 - \frac{\gamma - 1}{\gamma + 1}. \tag{8.41}$$

This relates to the temperature via $P = \rho k T/m$ where m is the mean particle mass. For a strong shock with $\gamma = 5/3$,

$$P_2 \to \frac{3}{4}\rho_1 v_1^2, \tag{8.42}$$

$$T_2 \to \frac{3}{16}\frac{m v_1^2}{k}. \tag{8.43}$$

This conversion of motion to heat can produce much hotter gas than radiative processes in the ISM and is discussed below in the context of supernova remnants.

Isothermal Shocks

The opposite extreme is when rapid radiative loss leads to effective cooling and a constant temperature across the shock. For typical cooling timescales in the WNM and HII regions of $\sim 10^4$ years and shock speeds of $\sim 10 \, \mathrm{km \, s^{-1}}$, the gas will cool after it travels ~ 0.1 pc. On scales larger than this, we can therefore consider the gas to be isothermal, $P = \rho c^2$, with the jump condition $T_1 = T_2$ to relate $P_1/\rho_1 = P_2/\rho_2$. It is left as an exercise at the end of the chapter to show that this leads to a quadratic equation in v_2,

$$v_1 v_2^2 - (c^2 + v_1^2)v_2 + c^2 v_1 = 0, \tag{8.44}$$

which has the roots $v_2 = v_1$ and $v_2 = c^2/v_1 = c/M_1$. The density contrast is then $\rho_2/\rho_1 = v_1/v_2 = M_1^2$ which increases without limit.

Figure 8.4 combines the two cases by showing the rapid jump in temperature in an adiabatic shock and its subsequent decline back to its original value as the cooling takes effect. The density jump is limited in the initial adiabatic shock but then continues to increase as energy is radiated away.

Having now determined the conditions across the shock, we can transform to different frames of reference. In particular, we can look at the relative motion of the gas up and downstream from the shock. Transforming to the upstream frame,

$$v_1 \to -v_{\mathrm{shock}}, \quad v_2 \to v_2 - v_{\mathrm{shock}}, \tag{8.45}$$

implies

$$\frac{v_2}{v_{\mathrm{shock}}} = 1 - \frac{1}{M_1^2}. \tag{8.46}$$

Fig. 8.4. The variation of
fluid properties in a strong
shock that is initially
adiabatic and then cools to
its original temperature. The
density contrast is initially
limited to 4 (for $\gamma = 5/3$)
but the gas continues to
compress as it radiates and
the initial temperature spike
subsides. The shading
represents the gas
temperature.

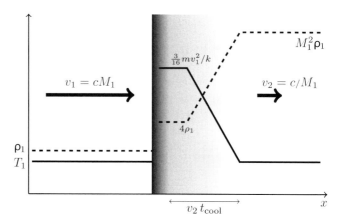

Fig. 8.4. The variation of
fluid properties in a strong
shock that is initially
adiabatic and then cools to
its original temperature. The
density contrast is initially
limited to 4 (for $\gamma = 5/3$)
but the gas continues to
compress as it radiates and
the initial temperature spike
subsides. The shading
represents the gas
temperature.

This implies that, in the case of strong isothermal shocks, all the post-shock gas is accelerated to the shock speed. We will now see the application of these ideas to the expansion of supernova remnants and HII regions.

8.7 Expanding Supernova Remnants

Supernovae may be caused by the detonation of a white dwarf through mass transfer from a binary companion or from the core-collapse of a massive OB star. The former are classified as Type Ia, the latter generally as Type II (very massive stars that blow off their outer hydrogen layers prior to collapse produce Types Ib and Ic). Core-collapse supernovae are more common and as their progenitors have short main sequence lifetimes, $t_{MS} = 3-30$ Myr, they are often associated with spiral arms and star-forming regions in the ISM. Regardless of origin, the impulsive energy input of a supernova dramatically affects its surroundings in a way that can be split into three physically distinct stages.

Initial Expansion

The first, short-lived, phase is free expansion at a constant speed. This holds until the mass of swept-up material is comparable to the mass of the ejecta. The ejecta from a typical supernova has kinetic energy $E_{SN} \sim 10^{44}$ J. This relates the speed to the ejected mass,

$$v_{ej} = 10^4 \left(\frac{M_{ej}}{1 \, M_\odot} \right)^{-1/2} \text{km s}^{-1}, \qquad (8.47)$$

and the radius in this phase is simply

$$R = v_{ej}t. \qquad (8.48)$$

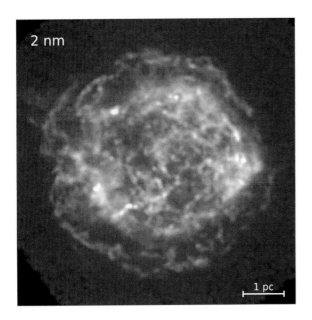

2 nm

1 pc

Fig. 8.5. X-ray image (3.3 − 10 keV) of the young Cassioppeia A supernova remnant, from data in the Chandra Supernova remnant catalog. The image is 7′ across.

If the density of the surrounding (pre-shock) material is ρ_1, then the free expansion will begin to slow down at a radius

$$R_{\text{free}} = \left(\frac{3M_{\text{ej}}}{4\pi\rho_1}\right)^{1/3}$$

$$\simeq 2.2 \left(\frac{M_{\text{ej}}}{1\,M_\odot}\right)\left(\frac{n_H}{10^6\,\text{m}^{-3}}\right)^{-1/3} \text{pc}. \tag{8.49}$$

This phase lasts for only a short time, $t_{\text{free}} = R_{\text{free}}/v_{\text{ej}} \simeq 200$ $(M_{\text{ej}}/1\,M_\odot)^{1/2}$ yr. An observational example is Cassiopea A, a remnant of a supernova estimated to have exploded about 300 years ago. It is one of the brightest radio sources in the sky and shines brightly across the electromagnetic spectrum through to the X-rays (Figure 8.5). The emission is from a combination of line radiation and the relativistic particles produced by Fermi acceleration in multiple shock waves.

Energy-Conserving Phase

The remnant continues to expand at speeds vastly greater than the sound speed of the ambient medium. The timescales are so short that energy is conserved and we can use the adiabatic jump conditions to describe the post-shock temperature and density,

$$\rho_2 = 4\rho_1,$$

$$T_2 = \frac{3}{16}\frac{mv^2}{k} = 2 \times 10^9 \left(\frac{v}{10^4\,\text{km s}^{-1}}\right)^2 \text{K}, \tag{8.50}$$

Fig. 8.6. Schematic of a supernova remnant during the blast-wave phase, where expansion into material of density ρ_1 shocks the gas (thick line) into a thin, dense, hot shell with density ρ_2. The shading represents the gas temperature.

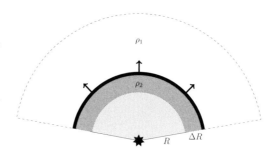

where we have assumed $\gamma = 5/3$ appropriate for expansion into atomic gas.

The shock creates a thin, compressed shell with width

$$\frac{\Delta R}{R} \simeq \frac{1}{3}\frac{\Delta \rho}{\rho} = \frac{1}{12}, \tag{8.51}$$

which is shown exaggerated in Figure 8.6.

The radius will grow with time but there is now an additional independent variable, ρ_1. Assuming power law dependencies, the requirement that the energy be conserved implies that the combination

$$E = E_{\rm SN} \propto \rho_1^a t^b R^c \tag{8.52}$$

is constant. This simple dimensional analysis argument has the unique solution $a = 1, b = -2, c = 5$, which implies that the radius varies as

$$R = R_{\rm scale}\left(\frac{E_{\rm SN}}{\rho_1}\right)^{1/5} t^{2/5}, \tag{8.53}$$

where $R_{\rm scale}$ is a scaling factor that depends on the initial conditions of this phase, or equivalently the end of the free-expansion phase above. This is known as the Sedov–Taylor blast solution based on their application to terrestrial explosions and, infamously by Taylor, to calculate the yield of the first nuclear weapon from time series photos of the Trinity test.

The expansion speed decreases with time as $v \propto t^{-3/5}$, as additional mass is set in motion but no more energy is added. Once the speed slows down to $v_{\rm blast} \sim 200\,{\rm km\ s^{-1}}$, the post-shock temperature drops to $\sim 10^6\,{\rm K}$ and hydrogen recombines. For the fiducial value of the ambient density, $n_e = 10^6\,{\rm m^{-3}}$, this occurs at $t_{\rm blast} \simeq 1.2 \times 10^5\,{\rm yr}$. The remnant radius is $R_{\rm blast} = 25\,{\rm pc}$, and the mass of swept up material is $1.6 \times 10^3\,M_\odot$.

The Crab nebula, shown in optical light in Figure 8.7, is a remnant of a supernova recorded by Chinese astronomers in 1054 and is in the early stages of the blast-wave phase. The central diffuse component results from synchrotron emission and the filamentary structures are

674 nm

1 pc

Fig. 8.7. Optical (DSS2 Red) image of the Crab nebula (M1), a supernova remnant in the energy-conserving phase of evolution. The image is 8′ across.

$H\alpha$ line emission. The nebula mass and expansion speed are 5 $M_\odot$ and 1500 km s^{-1}, respectively.

Momentum-Conserving Phase

As the temperature of the shocked gas drops, the cooling rate increases and the remnant transitions from adiabatic to fully radiative and effectively isothermal. Diffuse hot gas remains in the interior and the remnant continues to expand as momentum is conserved,

$$\rho_1 R^3 \frac{dR}{dt} = \text{constant.} \tag{8.54}$$

For a constant density ambient medium, the radius grows with time as $R \propto t^{1/4}$, and speed decreases as $t^{-3/4}$.

The deceleration of the remnant as material piles up is colloquially termed the snowplow phase. The Veil nebula in Cygnus is approaching this phase. It is about 8000 years old, 10 pc in radius, and has the appearance of a wispy network of gas filaments in a thin shell (Figure 8.8). This is actually a limb-brightened bubble glowing in the optical and ultraviolet due to collisionally excited lines produced at the shocked interface between the remnant and surrounding ISM.

Connecting the initial conditions for this phase to the final conditions, $\{R_{\text{blast}}, v_{\text{blast}}\}$, of the energy-conserving phase gives

$$R = R_{\text{blast}} + \left[4 v_{\text{blast}} R_{\text{blast}}^3 (t - t_{\text{blast}})\right]^{1/4}, \tag{8.55}$$

$$v = v_{\text{blast}} \left(\frac{R_{\text{blast}}}{R}\right)^3. \tag{8.56}$$

Fig. 8.8. Optical (DSS2 Red) image of the Veil nebula, an ~8000 yr old supernova remnant transitioning from the energy-conserving to momentum-conserving phase of evolution. The image is 3.2° ≃ 25 pc across.

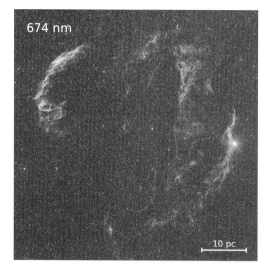

674 nm

10 pc

Figure 8.9 shows the three phases connected together. In practice, of course, the remnant smoothly transitions between each physical regime.

For our fiducial initial conditions, the supernova remnant expands to 73 pc and slows to 7 km s^{-1} at 10 Myr after the explosion. At this point, its speed is similar to the thermal speed in the warm neutral medium and the remnant will merge into the general ISM background. The mass of the remnant at this point is ~4 × 10^4 $M_\odot$ and its kinetic energy is ~2 × 10^{42} J which is just 2% of E_{SN}. This is sufficient, however, to stir up the ISM and maintain its turbulent motions which would otherwise dissipate rapidly. At the end of this chapter, we discuss the effects that supernovae have on the structure of the ISM through the creation and sustenance of extensive hot gas.

Fig. 8.9. The evolution of a supernova remnant through the freely expanding initial phase, followed by the adiabatic blast wave, and ending with the momentum-conserving "snowplow" phase. The solid line shows the radius and the dashed line the velocity.

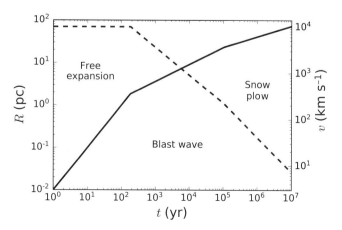

8.8 HII Region Evolution

Supernovae are relatively rare events, occurring about twice per century in the Galaxy. There are ~3×10^4 young clusters of O stars, however, and their associated HII regions also dramatically affects the ISM. Compared to the impulsive energy input from a supernova, HII regions provide a steady push, though still with a very large pressure jump since the ionization raises the temperature of the gas from tens of kelvin to about ten thousand kelvin. However, the very first stage in the evolution of an HII region, as a massive star lights up its surroundings, is an abrupt change in the ionization state of the gas.

Ionization Front

An ionization front is the boundary between neutral and ionized gas. As soon as the central star begins producing ionizing photons, it produces an HII region that, in ionization–recombination balance, would extend to the Strömgren radius defined in Equation 6.11,

$$R_S = \left(\frac{3 \dot{N}_{\text{ionize}}}{4 \pi \alpha_2 n_1^2} \right)^{1/3}, \tag{8.57}$$

where $\dot{N}_{\text{ionize}}$ is the production rate of ionizing photons from the central stars, α_2 is the total recombination rate to the Balmer level, and n_1 is the number density in the ambient medium. In practice, stars form in a dense core composed primarily of molecular hydrogen and some energy is required to dissociate them before ionization but, for illustrative purposes and ease of comparison with Chapter 6, we assume that the surrounding material is atomic hydrogen.

Figure 8.10 depicts the situation as the ionization front moves outwards. The speed of the front, V_{IF}, refers to the motion of the boundary between ionized and neutral gas and is not a dynamical

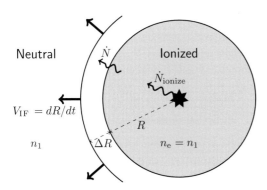

Fig. 8.10. The initial (non-dynamical) growth of a spherical ionization front into a neutral medium with hydrogen number density n_1. The central stars produce $\dot{N}_{\text{ionize}}$ photons per second, which decreases due to recombination within the ionized region to a rate $\dot{N}$ at the front.

movement of particles. Over a small time interval, Δt, the number of new ionizations is

$$\dot{N}\Delta t = 4\pi R^2 \Delta R n_1, \qquad (8.58)$$

where $\dot{N}$ is the ionization rate at the edge of the ionized region. This is reduced from the central production rate due to recombinations within the existing HII region,

$$\dot{N} = \dot{N}_{\text{ionize}} - \frac{4}{3}\pi R^3 \alpha_2 n_1^2. \qquad (8.59)$$

This then leads to a differential equation for the expansion rate,

$$3R^2\frac{dR}{dt} = \alpha_2 n_1 (R_{\text{S}}^3 - R^3), \qquad (8.60)$$

which has an exponential solution in R^3, or volume. The solution for radius is

$$R = R_{\text{S}}\left(1 - e^{-\alpha_2 n_1 t}\right)^{1/3}. \qquad (8.61)$$

The radius asymptotes toward R_0, with an e-folding time equal to the recombination timescale, $t_{\text{rec}} = 1/\alpha_2 n_1$. For an ionizing rate $\dot{N}_{\text{ionize}} = 10^{49}\,\text{s}^{-1}$, and ambient density $n_1 = 10^8\,\text{m}^{-3}$, $R_{\text{S}} = 3.1\,\text{pc}$, $t_{\text{rec}}1 = 1200\,\text{yr}$. The speed of the ionization front is very fast at first, $dR/dt \propto 1/R^2$, but approaches zero as $R \to R_{\text{S}}$. Eventually, we must therefore consider the motions of the gas particles themselves.

Pressure-Driven Expansion

The newly formed HII region has a temperature $T_{\text{e}} \simeq 10^4\,\text{K}$, which is over 100 times greater than its neutral surroundings and the ionization increases the overall number density by a factor of 2, $n_{\text{e}} + n_{\text{p}} = 2n_1$. It is therefore strongly over-pressured with respect to its surroundings and will expand supersonically. The consequent decrease in density reduces the number of recombinations per unit volume and the HII region therefore grows beyond R_0.

In this first stage of dynamical expansion, the neutral gas is strongly shocked and almost simultaneously ionized. We can describe the shock through the adiabatic jump conditions (Equations 8.33, 8.34),

$$\rho_1 v_1 = \rho_2 v_2,$$
$$\rho_1 v_1^2 = \rho_2 v_2^2 + P_2, \qquad (8.62)$$

where the subscript 2 refers to the post-shocked gas in the HII region and we have neglected the pressure term in the surroundings since it is overwhelmed by the HII region. This leads to a quadratic equation,

$$\left(\frac{\rho_2}{\rho_1}\right)^2 - \left(\frac{v_1}{c_2}\right)^2 \frac{\rho_2}{\rho_1} + \left(\frac{v_1}{c_2}\right)^2 = 0, \qquad (8.63)$$

where $c_2 = P_2/\rho_2$ is the sound speed in the ionized gas. Factoring out the density contrast, we arrive at

$$\left(\frac{\rho_2}{\rho_1} - \eta\right)^2 = \eta(\eta - 2), \qquad (8.64)$$

where

$$\eta = \frac{1}{2}\left(\frac{v_1}{c_2}\right)^2. \qquad (8.65)$$

The solution breaks down for $\eta < 2$, corresponding to when the speed of the gas entering the shock, v_1, decreases below $2c_2 = 18\,\text{km s}^{-1}$. The density contrast is then $\rho_2/\rho_1 = 2$. In our fiducial case, this occurs after $t_{\text{break}} \simeq 4t_{\text{rec}} = 5 \times 10^3$ yr at $R \simeq R_S = 3$ pc, so this is a relatively short-lived phase.

Thereafter, the ionization rate at the boundary drops below the flux of hydrogen atoms coming in and the ionization front breaks off from the dynamical shock wave. The gas is shocked and compressed but remains neutral. It cools rapidly through collisionally excited carbon and oxygen lines to approximately isothermal conditions before then being ionized by the stellar photons, as illustrated in Figure 8.11.

We therefore describe this second stage of dynamical expansion as an isothermal shock with density contrast $\rho_2/\rho_1 = M_1^2 = (V_s/c_1)^2$ where $V_s = dR/dt$ is the speed of the shock in the frame of the

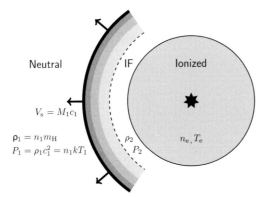

Fig. 8.11. The pressure-driven expansion of an HII region. The strong shock (solid line) creates a dense shell of neutral gas that cools down to the temperature of the surroundings and enters an ionization front (dotted line; transition labeled IF). The shading represents the gas temperature, which rises sharply in the adiabatic region of the strong shock, then cools to isothermal conditions before being reheated during ionization.

ambient medium (assumed stationary with respect to the star) and c_1 is the sound speed of the neutral gas. The post-shock pressure $P_2 = (\rho_2/\rho_1)P_1 = \rho_1 V_s^2$.

The HII region lies slightly further downstream with an electron density, n_e, that decreases as the radius increases due to ionization–recombination balance,

$$\dot{N}_{\text{ionize}} = \frac{4}{3}\pi R^3 \alpha_2 n_e^2 \equiv \frac{4}{3}\pi R_S^3 \alpha_2 n_1^2. \tag{8.66}$$

The expansion-driving pressure in the HII region is $P_{\text{HII}} = 2n_e k T_e$ where T_e is the electron temperature and the factor of 2 accounts for the contribution from the protons which are in energy equipartition. Equating this to P_2 then implies

$$n_1 m_H \left(\frac{dR}{dt}\right)^2 = 2k T_e n_1 \left(\frac{R_S}{R}\right)^{3/2}. \tag{8.67}$$

With initial conditions $R = R_S$ at $t = t_{\text{break}}$ for this second phase, the solution is

$$R = R_S \left[1 + \frac{(2k T_e/m_H)^{1/2}}{R_S}(t - t_{\text{break}})\right]^{4/7}. \tag{8.68}$$

The density decreases as $n_e \propto r^{-3/2} \propto t^{-6/7}$ and the expansion stalls when the density of the HII region drops to the point where the pressure matches its surroundings, $n_e = n_1 T_1/2T_e$.

Figure 8.12 plots the radius and velocity of the HII region against time for our nominal case. Pressure balance only occurs after $t \simeq 110\,\text{Myr}$ when the radius $R \simeq 100\,\text{pc}$, and the expansion speed has dropped to the sound speed of the ambient gas, $V_s = (n_1 k T_1/m_H)^{1/2}$. This situation is generally not reached in practice as the timescale is longer than the main sequence lifetime of the ionizing source.

Fig. 8.12. The pressure-driven expansion of an HII region with temperature $T_e = 10^4\,\text{K}$ into a cloud of atomic hydrogen with uniform density $n = 10^8\,\text{m}^{-3}$. The radius is shown as a solid line with scale on the left axis and velocity as a dashed line with scale on the right hand side. The gray area marks the region where the expansion stalls as the HII region pressure matches that of its surroundings.

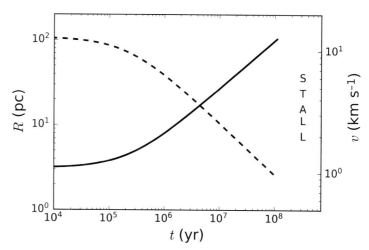

The total mass of the HII region at this point is $6 \times 10^4 \, M_{\odot}$, similar to that of supernova remnants. However, because the expansion speed is slow, its energy is about an order of magnitude lower and mainly in thermal motions, $\sim 2 \times 10^{41}$ J. The radiative energy summed over a typical O star main sequence lifetime, $E_{\mathrm{rad}} \sim \dot{N}_{\mathrm{ionize}} t \, h\nu \sim 2 \times 10^{44}$ J, is mainly re-radiated away through forbidden lines and the conversion to motion is very inefficient.

In practice, neither supernova nor HII regions expand into uniform density surroundings. Any inhomogeneities will affect the shape of the bubble as the ionization front will propagate further and the expansion will proceed faster through regions of lower density. Further, as the density of the bubble decreases below the ambient medium, Rayleigh–Taylor instabilities will result. These can be seen as the dense, dusty filaments at the edge of the Rosette HII region in Figure 6.1 and in the filamentary appearance of the Cassiopeia A and Veil supernova remnants (Figures 6.1 and 8.5). Nevertheless, the simplistic picture presented here illustrates the application of fluid equations to examining motions in the ISM and highlights many of the important physical concepts that explain various aspects of the ISM.

8.9 The Hot Ionized Medium

Early X-ray surveys revealed, in addition to many compact Galactic and extragalactic sources, diffuse, low-intensity emission along the Galactic plane. The implication of a very hot component of the ISM was confirmed through UV spectroscopy which detected lines from OVI and other highly ionized species. To remove an electron from the highly positive O^{4+} potential requires a very high energy, $E = 114\,\mathrm{eV}$, that is well beyond typical radiative energies in the ISM and indicates high-speed collisions in gas with temperatures $T \sim E/k > 10^6\,\mathrm{K}$. Their broad linewidths are consistent with the large thermal velocity dispersion in such hot gas.

This phase of the ISM is termed the **hot ionized medium** (HIM), also sometimes called coronal gas because of its similarity in temperature with the solar corona. The HIM is created in strong adiabatic shocks at speeds $\gtrsim 200\,\mathrm{km\,s^{-1}}$ (Equation 8.43). Late stages of massive stellar evolution can result in high-speed winds of this magnitude but the affected volume is small. Supernovae heat much larger regions and have an even greater collective impact.

Super-hot, shock-produced, gas will expand until its pressure drops to that of the other phases of the ISM. At this point, its number density, $n \propto 1/T \sim 10^4\,\mathrm{m^{-3}}$, is about 100 times lower than that in the WIM and WNM and 10^4 times lower than in the CNM. The rate of two-body radiative processes including recombination, collisional excitation, and bremsstrahlung scales as n^2 and is extremely low compared to these other phases. Consequently, the HIM emits weakly, cools slowly, and lasts for a long time.

In the discussion on supernova evolution above, the energy-conserving phase ends about 10^5 yr after the explosion when the remnant is 25 pc in radius. At this point, the gas at the shock front is at 10^6 K and much hotter at the remnant center. The remnant then enters the snowplow phase where it cools off, slows down, and eventually merges into the background at about 10 Myr. For demonstrative purposes here, we take the geometric mean, $t = 1$ Myr, as a representative timescale of the HIM-producing phase. From Figure 8.9, the radius of the hot bubble is then $R_{\rm hot} \simeq 50$ pc.

The supernova rate in the Galaxy, S, is estimated to be about two per century (most are far away and obscured by dust, but we appear to be overdue), which implies a total number of hot supernova remnants, $St_{\rm hot} \simeq 2 \times 10^4$, each with a volume $V_{\rm hot} = 4/3\pi R_{\rm hot}^3$. We can therefore express the total volume of hot gas as

$$V_{\rm HIM} \simeq 10^{10} \left(\frac{R_{\rm hot}}{50\,{\rm pc}} \right)^3 \left(\frac{S}{0.02\,{\rm yr}^{-1}} \right) \left(\frac{t_{\rm hot}}{1\,{\rm Myr}} \right) {\rm pc}^3. \qquad (8.69)$$

To put this into perspective, consider the volume of the Galaxy as a cylinder with radius out to the edge where massive stars are found $R_{\rm OB} = 10$ kpc, and a thickness of the dense star-forming gas $H \simeq 200$ pc (see Chapter 10),

$$V_{\rm tot} = \pi R_{\rm OB}^2 H = 6 \times 10^{10}\,{\rm pc}^3. \qquad (8.70)$$

These are very crude estimates but suggest that the HIM occupies about 20% of the total volume of the ISM within the Galactic disk where supernova-producing stars are found. At this level of porosity or filling factor, $\phi = V_{\rm HIM}/V_{\rm tot} \simeq 0.2$, the individual bubbles of hot gas are likely to merge with each other. Consider a spherical region with radius $2R_{\rm hot}$. Two bubbles with radius $R_{\rm hot}$ cannot fit into this region without overlapping with each other. However, the total volume is $8V_{\rm hot}$ and, on average, the HIM occupies $8V_{\rm hot}\phi$ so the mean number of individual bubbles is $\mu = 8\phi \simeq 1.6$. Using Poisson statistics, the probability that this region contains no bubbles of hot gas is $e^{-\mu} = 0.2$, and one bubble is $\mu e^{-\mu} = 0.3$. Thus the likelihood of two or more bubbles is 0.5 and these will necessarily merge together.

These probabilities depend exponentially on the filling fraction, ϕ, and are therefore highly sensitive to the evolution (radius and lifetime) of the hot supernova phase. The weak and diffuse nature of the emission from the HIM complicates observational tests, but detailed models

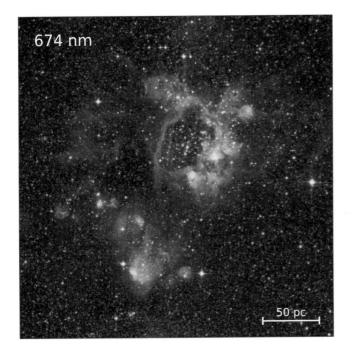

674 nm

50 pc

Fig. 8.13. Optical (DSS2 Red) image of the N44 nebula in the Large Magellanic Cloud. The nebulosity is collisionally excited line emission from gas that has been shocked by previous supernovae and irradiated by a new generation of massive stars. The combined effect has blown a huge hole, or superbubble, in the galaxy. The image is $0.3° \simeq 260\,\mathrm{pc}$ across.

and numerical simulations agree that supernova-driven shocks produce large volumes of hot gas in the ISM and that remnants can congregate together, exchanging heat and maintaining the HIM. Massive stars tend to form in groups and produce concentrated bursts of supernovae closely spaced in location and time. These create the superbubbles of hot gas that can be seen in the Large Magellanic Cloud (Figure 8.13). These revisions to the large-scale structure of the ISM are the subject of Chapter 10, and the effect of superbubbles on galaxy evolution is described further in Chapter 11.

The pervasiveness of the HIM regulates the pressure in the ISM, which is high enough to allow the cold phase of atomic gas to exist. This in turn leads to regions of high dust extinction and shielded environments where molecules can form and cool the gas still further to sufficiently high densities for self-gravity to turn the ISM into stars.

Notes

For more detail on astrophysical applications of fluid dynamics see the textbook by Clarke and Carswell (2014). A gallery of spectacular Chandra satellite X-ray images of supernova remnants is at hea-www.harvard.edu/ChandraSNR. The addition of the HIM to the pressure-regulated WNM and CNM led to the idea of a three-phase medium and one of the most highly cited papers on the ISM by McKee and Ostriker (1977).

Questions

1. Derive Equation 8.15 describing the growth of small density perturbations in a uniform, self-gravitating fluid.

2. Derive the expression for the gravitational potential, $W = \frac{3}{5}GM^2/R$, for a uniform sphere used in the section on the virial theorem.

3. Fill in the intermediate algebraic steps describing the pressure jump of an adiabatic shock (Equation 8.41) and the quadratic equation relating the speeds before and after an isothermal shock (Equation 8.44).

4a. Verify the dimensional analysis for the blast wave solution in Equation 8.53.

4b. Look up online images of supernova remnants. Note the scale and wavelength and estimate where they lie in the evolutionary sequence described in Section 8.7.

5. Consider the evolution of an HII region under pressure-driven expansion in Figure 8.12. Referring back to Chapter 6, discuss how this manifests itself in the radio SED of bremsstrahlung emission. What is the approximate proportion of HII regions that you would expect to see in the ultra-compact phase, $R < 1$ pc, compared to objects like the Rosette nebula and how might you check this? (It turns out that there are many more ultra-compact HII regions than expected, first shown by Wood and Churchwell (1989). Various explanations have been made to explain why young HII regions do not expand as fast as our simple model predicts.)

Chapter 9
Star Formation

The ISM is constantly forming stars. Infrared observations show young stars embedded in molecular clouds. Studies of the dense cores in these clouds show how stars are born and, by inference, tell us the story of our own origins.

Star formation is a challenging subject on account of the huge range of scales involved, ranging from a few tenths of a parsec, $\sim 10^{16}$ m, for a molecular core to stellar radii, $\sim 10^9$ m. The corresponding range of densities spans an astonishing 21 orders of magnitude, from $\sim 10^9$ hydrogen molecules per m^3 for the average density of a core to $\sim 10^{30}$ protons per m^3 for the average density of the Sun. This chapter describes how the inescapable force of gravity produces such tremendous compression in the ISM, the nature of the resulting protostars, and why the birth of stars inherently provides the conditions for planetary systems to form.

9.1 Gravitational Collapse

The Jeans criterion gives a mass scale above which a thermally supported cloud would become unstable and gravitationally collapse. The Jeans mass decreases as the gas becomes denser and colder because the gravitational force becomes stronger and thermal pressure support proportionally weaker. If we consider a cloud of pure molecular hydrogen, the prescription in Equation 8.19 can be reformulated into physical units as

$$M_{\rm J} = 3.9 \left(\frac{n_{\rm H_2}}{10^{10}\,{\rm m^{-3}}} \right)^{-1/2} \left(\frac{T}{10\,{\rm K}} \right)^{3/2} M_\odot. \qquad (9.1)$$

This shows that the dense, cold cores in molecular clouds are the most viable sites for gravitational collapse at stellar mass scales. The lower density, warmer CNM with $n_{\rm H} \sim 10^8\,{\rm m^{-3}}, T \sim 100\,{\rm K}$, does not become gravitationally unstable until much higher masses, $\sim 10^3 \, M_\odot$, and orders of magnitude more for the more rarefied, hotter WNM, WIM, and HIM. Indeed, gravitational collapse and protostars are observed in molecular cloud cores, generally through observations at infrared wavelengths that pass through the high dust columns and at millimeter wavelengths that detect molecular rotational transitions.

In the absence of any support, the collapse can be described as a free-fall with acceleration determined by the gravitational force,

$$m \frac{d^2 r}{dt^2} = -\frac{GMm}{r^2}, \qquad (9.2)$$

where m is a test particle on the outer shell of a spherical core with mass $M \gg m$. This can be simplified by noting $d^2 r/dt^2 = \frac{1}{2} dv^2/dr$ and integrating once with the boundary condition $v = 0$ at the outer radius $r = r_0$, to obtain

$$\frac{dr}{dt} = -\left[2GM\left(\frac{1}{r} - \frac{1}{r_0}\right)\right]^{1/2}, \qquad (9.3)$$

where the negative square root is chosen because the core is collapsing. The substitution $r = r_0 \sin^2 \theta$ then leads to an expression for the test particle to move from the outer radius to the center,

$$t_{\rm ff} = \left(\frac{3\pi}{32 G \rho}\right)^{1/2} = 1.2 \left(\frac{n_{H_2}}{10^{10} \, {\rm m}^{-3}}\right)^{-1/2} {\rm Myr}, \qquad (9.4)$$

where $\rho = 3M/4\pi r_0^3 = n_{H_2} m_{H_2}$. This is known as the gravitational free-fall time and shows that the formation of a solar-mass star happens within the first 0.01% of their $\sim 10\,{\rm Gyr}$ main sequence lifetime. There are plenty of stars, however, and many instances where we can observe this rare and short-lived phase through structural and kinematic signatures in the dense gas.

9.2 The Bonnor–Ebert Sphere

The free-fall time is a useful guide to the scales involved in the collapse of a dense molecular core to form a star but, in practice, gravity is resisted by other forces. This can slow down collapse or allow stable cores to exist. A spherical core in balance between self-gravity and thermal pressure can be described by expressing the fluid equations 8.5, 8.6, 8.9 in spherical coordinates as follows,

$$\nabla^2 \phi = \frac{1}{r^2} \frac{d}{dr}\left(r^2 \frac{d\phi}{dr}\right) = 4\pi G \rho,$$

$$\frac{dP}{dr} = -\rho \frac{d\phi}{dr}, \qquad (9.5)$$

$$P = \rho c^2.$$

The first line is Poisson's equation, the second expresses hydrostatic equilibrium, and the third is the equation of state where, for simplicity, we assume an isothermal temperature T with corresponding sound speed $c = (kT/m_{H_2})^{1/2}$.

These equations have a simple power law solution. If we try $\rho = \rho_0 r^p$, then $P = \rho_0 c^2 r^p$, $d\phi/dr = -(dP/dr)/\rho = -pc^2/r$, and Poisson's equation becomes $pc^2/r^2 = -4\pi G \rho_0 r^p$ which implies $p = -2$ and the solution

$$\rho = \frac{c^2}{2\pi G} r^{-2}. \qquad (9.6)$$

This is called the **singular isothermal sphere** (SIS) and is theoretically appealing because of its analytic simplicity. It allowed progress in developing models toward testable predictions that matched many of the early observations of dense molecular cores. It is physically unrealistic, however, in that the density and mass each increase without limit on small and large scales respectively. Indeed, as technology allowed higher resolution observations at millimeter wavelengths, it became apparent that core densities reached a plateau toward the center.

For the more general case of a finite central density, ρ_{cen} at $r = 0$, we must numerically integrate the equations. These can be simplified by combining the second and third equations in 9.5 to

$$\rho = \rho_{cen} e^{-\phi/c^2}, \qquad (9.7)$$

and using normalized variables, $x = (4\pi G \rho_{cen}/c^2)^{1/2} r$, $y = \phi/c^2$, to produce the dimensionless equation,

$$\frac{1}{x^2} \frac{d}{dx}\left(x^2 \frac{dy}{dx}\right) = e^{-y}. \qquad (9.8)$$

This is the Emden–Chandrasekhar equation and it can be numerically solved (steps for how to do this are given in an exercise at the end of this chapter). Profiles for different central densities are shown in Figure 9.1 and are overlaid with the SIS which represents the limiting case of infinite central density.

The numerical solutions address the density singularity at the core center but they still have arbitrarily large masses at large scales since they have the same power law behavior as the SIS, $\rho \propto 1/r^2$. In practice, however, the density does not decrease to zero but the core merges into the general ISM. We can model this by including an ambient pressure that sets a constraint on the outer radius, $n_{amb} = P_{amb}/kT$. As we have discussed throughout the book, the ISM is in approximate thermal pressure equilibrium with $nkT \simeq 10^{-13}$ Pa. For the case here, where $T = 10$ K, this corresponds to $n_{amb} = 8 \times 10^8$ m^{-3}, shown by

Fig. 9.1. Density profiles for Bonnor–Ebert spheres with different central H_2 densities, ρ_{cen}/m_{H_2}, at a temperature of 10 K. The black dashed line shows the limiting case of the singular isothermal sphere. The horizontal dashed line shows a representative ambient density that determines the labeled stable and unstable regimes discussed in the text.

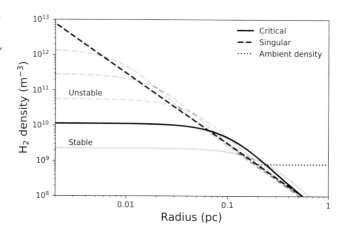

the horizontal dashed line in Figure 9.1. This sets a finite boundary at ~0.2 pc for the core radius and therefore a finite mass.

For a given mass, the equilibrium core radius depends on the central density. Cores with high central concentrations have slightly smaller radii than those with lower central densities. The most centrally concentrated case is the SIS which matches the density of the ambient medium at a radius

$$R_{SIS} = \frac{c}{(2\pi G \rho_{amb})^{1/2}}. \qquad (9.9)$$

The mass contained within this region is

$$M_{SIS} = \int_0^{R_{SIS}} 4\pi r^2 \rho \, dr = \left(\frac{2}{\pi G^3 \rho_{amb}}\right)^{1/2} c^3, \qquad (9.10)$$

where $\rho_{amb} = n_{amb} m_{H_2}$.

For the more general case, the mass is related to the dimensionless variables,

$$M_{BE} = \left(\frac{c^3}{4\pi G^3 \rho_{amb}}\right)^{1/2} \left[\left(\frac{\rho_{cen}}{\rho_{amb}}\right)^{-1/2} \int_0^{x_{amb}} x^2 e^{-y} \, dx\right], \qquad (9.11)$$

where the upper limit on the integral is set by the ambient density constraint, $y(x_{amb}) = c^2 \ln(\rho_{cen}/\rho_{amb})$. The term in square brackets depends only on the core density contrast, ρ_{cen}/ρ_{amb}, and can be evaluated numerically. The mass is plotted as a function of the density contrast in Figure 9.2 (for the same values of temperature and ambient density in the previous Figure 9.1). As the central density increases, the mass initially also increases but eventually the core shrinks, the volume decreases, and the enclosed mass drops. It asymptotes to M_{SIS} at very high density contrast. The maximum value, known as the **Bonnor–Ebert mass**, is 50% larger than M_{SIS} and occurs at $\rho_{cen}/\rho_{amb} = 14$.

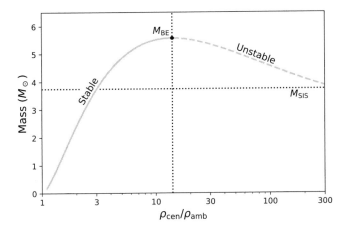

Fig. 9.2. The mass of the Bonnor–Ebert density profiles in Figure 9.1 for different values of the ratio between the central and ambient density, ρ_{cen}/ρ_{amb}. The dashed lines show the pressure ratio of 14 at the maximum mass, M_{BE}, and the mass of a singular isothermal sphere, M_{SIS}.

The density profiles and masses shown in these two plots are mathematical solutions to the equilibrium set up in Equations 9.5 but we need to consider the physics to determine their stability. If we imagine a core growing from the ambient medium, it will start from zero mass at $\rho_{cen}/\rho_{amb} = 1$. As its mass increases, the core will settle into ever more concentrated profiles until its central density reaches the maximum at $\rho_{cen}/\rho_{amb} = 14$. Beyond this point, it cannot gain additional mass and remain in equilibrium because it will then be so dense that self-gravity overcomes thermal pressure support and the core collapses. Thus the left hand side of Figure 9.2 is stable and the right side unstable. The corresponding profiles are shown by the solid and dashed gray lines in Figure 9.1.

Alternatively, we can imagine a core with the maximum stable mass in an ambient medium that changes. If external forces change the pressure and therefore the ambient density, then the core will become unstable and collapse. This is the essence behind the idea of **triggered star formation**, by which the impulsive force of an expanding HII region or supernova remnant can induce the collapse of marginally stable cores in neighboring molecular clouds.

Observations of isolated, dark clouds known as Bok globules provide an excellent test of these calculations. These clouds are dense enough to block light in the optical but become transparent in the near-infrared, an example being Barnard 68 in Figure 4.1. The visual extinction, which is proportional to the total column density, can be derived from star counts and reddening measurements as described in Chapter 4. The radially averaged profile in Figure 9.3 shows a flat center and steep fall off that is very well matched by a Bonnor–Ebert profile integrated along the line of sight. The inferred central density contrast is

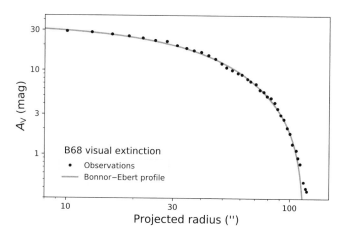

Fig. 9.3. The visual extinction, A_V, in magnitudes versus radial separation for the Barnard 68 dark cloud. The dots show the observed A_V derived from star counts in the optical and near-infrared. The solid gray line shows the integral of a Bonnor–Ebert density profile for different lines of sight across the spherical core.

close to the critical value of 14 and shows that the core is on the brink of instability. If it were to gain additional mass or be externally perturbed, it would undergo collapse.

Note that the Bonnor–Ebert mass scales with the temperature and density as $T^{3/2}\rho^{-1/2}$ in the same way as the Jeans mass and has a similar magnitude. The derivation here is a more formal calculation with well-defined boundary conditions that avoid the "swindle" alluded to in Chapter 8. Whereas these considerations apply for isothermal pressure support against gravity, they can be generalized to adiabatic equations of state. Turbulent pressure can also be an important source of support. To the extent that it can be modeled as an isotropic force with $P = \rho\sigma^2$ where σ is the one-dimensional turbulent velocity dispersion, a limiting mass can be inferred with σ replacing c above.

9.3 Observations of Core Collapse

The gravitational collapse of a cloud core, though fast by astronomical standards, is far too slow to produce a detectable signature through a change in radius or density profile. Nevertheless, we can observe the associated motions through Doppler shifts of molecular spectral lines. A collapsing core will have a greater velocity dispersion than a static core but, because we only measure projected motions along the line of sight, we cannot distinguish collapse from expansion via the dispersion alone. The trick lies in considering observations of an optically thick line such that we effectively only see the motion of the front side of the core relative to its center. The situation is illustrated in Figure 9.4 where a spherical core is represented, along a single line of sight, by two plane

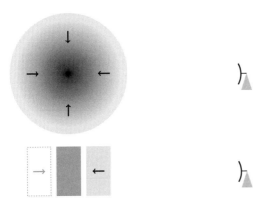

Fig. 9.4. A collapsing core
with a central density
gradient (top left) observed
by a radio telescope (top
right) can be approximated
by a slab model (bottom
left) for an optically thick
line. The observed emission
comes predominantly from
the center and the layer
closest to the observer and
the furthest layer is shown
dashed as it is not included
in the model. The arrows
represent the gas flow
relative to the core center.

layers (a third layer furthest from the observer is ignored as its emission
is mostly absorbed by the center).

This simplified model allows us to write an expression for the
observed spectrum. Since we are only interested in the radiative transfer,
the relevant parameters are the excitation temperatures and optical
depths of the front and back layers as T_f, τ_f and T_b, τ_b respectively. The
specific intensity is then the superposition of the front layer and the back,
attenuated by the front,

$$I_\nu = B_\nu(T_f)\,(1 - e^{-\tau_f}) + B_\nu(T_b)\,(1 - e^{-\tau_b})\,e^{-\tau_f}. \qquad (9.12)$$

For many cases, the density and temperature decrease away from the
center and therefore the excitation temperature is lower in its outer parts,
$T_b > T_f$. This implies that the front layer will absorb some of the
emission from the back, a phenomenon known as **self-absorption**.

Now moving into velocity space relative to the core center (back
layer), we assume that the optical depth in each layer is distributed as
a Gaussian with uniform velocity dispersion σ and peak values τ_{f0}, τ_{b0}
such that

$$\tau_f = \tau_{f0}\, e^{-(v - v_{in})^2/2\sigma^2},$$
$$\tau_b = \tau_{b0}\, e^{-v^2/2\sigma^2}. \qquad (9.13)$$

Here, v_{in} is the infall speed of the collapsing core with the front layer
moving from the observer toward the back layer.

The model has a total of six parameters: the excitation temperatures
determine the flux scale and the dispersion sets the velocity scale. The
general shape of the spectrum is primarily determined by the front layer
optical depth and ratio of infall speed to dispersion. Figure 9.5 shows

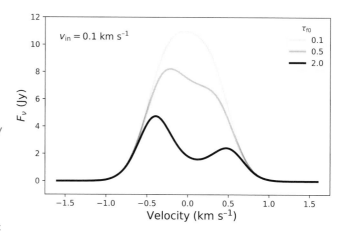

Fig. 9.5. The variation of the spectral profile from a two-layer model as the optical depth of the front (infall) layer increases from thin to thick. This model is for CS 2–1 at 98 GHz, with an optical depth of 2 in the back layer, a uniform velocity dispersion of 0.5 km s^{-1}, infall speed of 0.1 km s^{-1}, and excitation temperatures of 5 K and 20 K in the front and back layers respectively. The flux density is determined from the specific intensity assuming a resolution of 10″.

how the spectral profile for a fixed v_{in} changes from near-Gaussian for $\tau_{f0} \ll 1$ to double-peaked for $\tau_{f0} > 1$ as the front layer absorbs an increasing amount of the emission from the back layer.

The low-velocity peak is greater than the high-velocity peak because the front layer is moving away from the observer, $v_{in} > 0$, and therefore predominantly absorbs the redshifted side of the emission from the back layer. The stronger the absorption, the stronger the effect on the line profile. In general, astronomers like to observe optically thin lines since we can then count the number of emitting particles and determine column densities. This is an unusual case where we actually learn more from observing an optically thick line.

Figure 9.6 plots a series of profiles for varying infall speed and fixed front layer optical depth. The profile is symmetric for a static core, $v_{in} = 0$, but becomes increasingly asymmetric as v_{in} increases in magnitude. The left panel shows a collapsing core, $v_{in} > 0$, and the right panel shows an expanding core with $v_{in} < 0$ (front layer moves toward the observer). Infall motions have a telltale signature of redshifted absorption. More generally, the sensitivity to v_{in} shows that we can distinguish relative motions in a core and determine how fast they are.

Blue-shifted absorption, corresponding to expansion, is sometimes seen in the spectra of massive stars due to a wind blowing away from the photosphere and is known as a P-Cygni profile. Thus the signature of a collapsing core is referred to as an **inverse P-Cygni profile**.

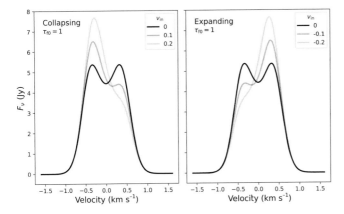

Fig. 9.6. The sensitivity of the spectral profile to the relative speed of the front layer. The model parameters are otherwise the same as Figure 9.5, but with a fixed front layer optical depth of 1. The left hand panel shows a collapsing core with a positive infall speed (away from the observer) and the right hand panel shows the opposite case where the front layer is moving toward the observer as if the core were expanding.

9.4 Observations of Protostars

From the expressions for the free-fall timescale and maximum stable core mass, we can derive an upper limit to the mass infall rate,

$$\dot{M}_{in} = \frac{1.5 M_{SIS}}{t_{ff}} \approx 2.2 \frac{c^3}{G}. \tag{9.14}$$

There is no density dependence because M_{SIS} and t_{ff} both scale in the same way as $\rho^{-1/2}$. Physically, denser cores become unstable at lower masses but have faster collapse times. For a 10 K core supported only by thermal pressure, $\dot{M}_{in} \approx 4 \times 10^{-6}$ $M_\odot$/yr. This is high enough that the release of gravitational energy, as gas falls from core to stellar scales, provides significant **accretion luminosity**,

$$L_{acc} = -GM\dot{M}_{in} \left(\frac{1}{R_{core}} - \frac{1}{R_*} \right)$$

$$\approx 9.3 \left(\frac{M}{1\,M_\odot} \right) \left(\frac{\dot{M}_{in}}{10^{-6}\,M_\odot\,yr^{-1}} \right) \left(\frac{R_*}{3\,R_\odot} \right)^{-1} L_\odot. \tag{9.15}$$

The core radius is so much larger than the stellar scale that the initial potential energy is effectively zero. The normalization of the protostellar radius, R_*, is a representative value for solar-mass stars at early times.

This order-of-magnitude calculation shows that a protostar is detectable well before it begins nuclear fusion. Due to the high dust extinction of the surrounding core, the very earliest phase of star formation is only visible at $\lambda > 100$ μm but, within $\sim 10^5$ yr after collapse, protostars become detectable in the near- and mid-infrared (Figure 9.7).

As the core is used up or otherwise dispersed, both the dust emission and extinction decrease. The protostar becomes increasingly visible at shorter wavelengths and the longer wavelength emission decreases. This is often referred to as the SED shifting toward the blue, where the

Fig. 9.7. Image of the Ophiuchus star-forming region at 24 μm, 0.84 × 0.70 degrees in size, produced from observations with the Spitzer space telescope. Numerous pre-main sequence stars are detected against nebulous warm dust emission from the surrounding molecular cloud. Courtesy NASA/JPL-Caltech.

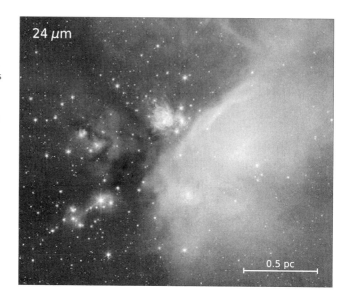

"blue" here is a catch-all term to denote *relatively* short wavelengths. The evolutionary state of a protostar is generally classified based on its infrared SED, but millimeter wavelength imaging of its surrounding envelope and disk (see below) is increasingly used in addition.

Once the collapse phase ends and the protostar has reached its final mass, it is known as a **pre-main sequence star**. Though hydrostatically supported, it slowly contracts through the loss of gravitational energy by radiation on a **Kelvin–Helmholtz timescale**,

$$t_{\mathrm{KH}} = \frac{GM_*^2}{R_* L_*} \simeq 31.1 \left(\frac{M_*}{M_\odot}\right)^2 \left(\frac{R_*}{R_\odot}\right)^{-1} \left(\frac{L_*}{L_\odot}\right)^{-1} \mathrm{Myr}. \quad (9.16)$$

For most stars, this is much longer than the dynamical timescale for core collapse, $t_{\mathrm{ff}} \sim 1\,\mathrm{Myr}$, and thus pre-main sequence stars can be observed at optical wavelengths after they emerge from their dusty birthplace. Indeed, **T Tauri stars** with $0.1\,M_\odot < M_* < 2\,M_\odot$ were first characterized in the 1940s when studies of the ISM were in their infancy, and this remains a common alternative name for pre-main sequence stars for historical reasons. Their intermediate mass counterparts, $2\,M_\odot < M_* < 8\,M_\odot$, are known as **Herbig Ae/Be stars**, where the A and B refer to the stellar spectral types and the e denotes the photospheric emission lines that were used to identify their youth. Since more massive stars are much more luminous, t_{KH} decreases with mass and ultimately becomes lower than t_{ff} for $M_* \gtrsim 8\,M_\odot$. There is therefore no pre-main sequence phase for such massive stars.

Pre-main sequence stars are initially bright due to their large sizes but grow fainter as they become smaller, while maintaining approximately constant temperature. The first stages of their evolution on the **Hertzsprung–Russell diagram** is a near vertical line known as the **Hayashi track**. Ultimately their central temperatures increase to $\sim 10^7$ K where hydrogen fusion begins and they become main sequence stars.

As is apparent in Figure 9.7, stars tend to form in groups, or **protostellar clusters**. The range of luminosities here is due partly to different evolutionary states but mostly to a range of stellar masses and therefore relates to the way in which stars form. The mass, size, and timescale of a collapsing core all vary with the inverse square root of density, which leads to the idea of hierarchical fragmentation noted in Chapter 8. However, this simple picture assumes that the gas remains isothermal or equivalently that the gravitational energy is radiated away. Because smaller fragments are denser, they are more optically thick so radiation only escapes from their surface at a rate bounded by a perfect (blackbody) emitter, σT^4 per unit area. If the accretion luminosity is greater than this, then the fragment will heat up and its increasing pressure will eventually resist collapse. Thus we have the condition

$$\frac{GM^2}{Rt_{\mathrm{ff}}} < 4\pi R^2 \epsilon \sigma T^4, \tag{9.17}$$

where the free-fall timescale is defined in Equation 9.4 and ϵ is the radiative efficiency. Substituting $\rho = 3M/4\pi R^3$ and the expression for the Jeans length in Equation 8.18 results in an expression between the mass and temperature,

$$M > 0.01 \epsilon^{-1/2} T^{1/4} M_\odot. \tag{9.18}$$

A slightly different pre-factor is obtained if we use the Bonnor–Ebert scale instead of Jeans. This basic reasoning shows that fragmentation in a cold molecular cloud, $T = 10$ K, with efficient radiative cooling may proceed down to planetary scales, $M \simeq 0.02\,M_\odot \simeq 20\,M_{\mathrm{Jup}}$. If, instead, we consider fragmentation as a protostar forms with $T \sim 10^3$ K and a much lower radiative efficiency, $\epsilon = 0.1$, the lower mass limit is an order of magnitude higher, $M \simeq 0.2\,M_\odot$. Given these considerations, we might expect a soft lower limit with the number of stars declining over this mass range.

There is no corresponding limit to the upper mass scale from considerations of ISM structure. Observations suggest that stellar masses can exceed $\sim 200\,M_\odot$, though such behemoths are very rare.

By studying a protostellar cluster, we can determine the number of stars, ΔN, in a given mass bin, ΔM. This results in what is known as the **initial mass function** (IMF), often described in differential form, dN/dM, or its dimensionless variant, $M dN/dM = dN/d\ln M$. The "initial" is here because stellar lifetimes depend on mass so the mass

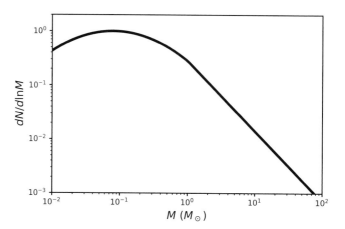

function (or distribution) evolves with time. The most massive stars
are extremely rare. There are about 200 stars with masses between
1 and 2 $M_\odot$ for every one between 50 and 100 $M_\odot$. Above a solar
mass, the IMF is described well by a power law, $dN/d \ln M \propto M^{-\Gamma}$,
where $\Gamma = 1.35$ is known as the Salpeter IMF slope after the first
person to describe it. Low-mass stars, below the fragmentation limit
described above, are also rare. The IMF turns over below about 0.2 $M_\odot$,
corresponding to spectral type M4. The exact form at lower masses is
hard to characterize due to the faintness of such objects, but the general
form can be approximated as a Gaussian in the logarithm of the mass.
Figure 9.8 plots a representative distribution that matches observations
of nearby star-forming regions.

Such a log-normal distribution is the natural outcome for a complex
system (star formation) in which the end result (stellar mass) is the product of the efficiencies of many independent processes. The logarithm of
the quantity is then the sum of a series of random variables and, by the
central limit theorem, is normally distributed.

Allowing for the observational uncertainties at the low-mass end and
small number statistics at the high-mass end, the observed distribution
of stellar masses is remarkably uniform across different stellar clusters,
young and old, and with the general field population in the Galaxy (after
taking stellar evolution into account). This suggests that star formation
is insensitive to large-scale environment. The majority of stars, by both
number and integrated mass, lie within a factor of 3 of $\sim$0.5 $M_\odot$. The
existence of this preferred mass scale may be due to self-regulation in
the star formation process, but a simpler explanation is that it arises
from fundamental properties of the ISM, namely the low temperatures
in molecular clouds that allow for self-gravity to overcome thermal
pressure at solar-mass scales.

9.5 The Rate and Efficiency of Star Formation

The **star formation rate** (SFR) is the total stellar mass formed in a given time interval. It can be defined for an individual star-forming region or for the Galaxy as a whole. For a molecular cloud, infrared surveys, such as shown in Figure 9.7, are used to count the number of protostars and pre-main sequence stars and determine their evolutionary state, from which we can estimate masses and ages. By restricting the count to an age range $[t_0, t_1]$, we can determine the SFR,

$$\dot{M}_* = \frac{\sum M_*}{t_1 - t_0}. \tag{9.19}$$

The accuracy to which we can determine masses and ages of an individual young star is poor, but the SFR can be estimated reasonably well when averaged over a large cluster. We can then determine the Galactic SFR through a census of all nearby ($\lesssim 1$ kpc) star-forming regions and extrapolating based on models of their spatial distribution.

An alternative approach is to observe the most massive stars, with spectral types OB. These have short main sequence lifetimes measured in Myr rather than Gyr for solar-mass stars so their numbers translate directly to their formation rate, and the associated HII regions are bright enough to be observed at much greater distances, $\gg 1$ kpc, than low-mass stars. As described in Chapter 6, optical or (less extincted) infrared spectroscopy of recombination lines and radio continuum observations relate to the ionization rate and therefore stellar mass. The overall SFR is then

$$\dot{M}_* = \frac{\sum M_{\rm OB}}{f \langle t_{\rm OB} \rangle}, \tag{9.20}$$

where f is the mass fraction of massive stars, and is determined by integrating the IMF, $\xi(M) = dN/d\ln M$,

$$f = \frac{\int_{M_{\rm OB, min}}^{M_{\rm max}} M\xi(M) d\ln M}{\int_{M_{\rm min}}^{M_{\rm max}} M\xi(M) d\ln M}. \tag{9.21}$$

The mean lifetime, $\langle t_{\rm OB} \rangle$, is similarly determined as an IMF weighted average.

A meta-analysis of many studies finds that the overall Galactic SFR is $1.7 \, M_\odot \, {\rm yr}^{-1}$. This is about a factor of 3 times lower than the rate required to build up the Galaxy with total stellar mass $\sim 6 \times 10^{10} \, M_\odot$ in a Hubble time, 13.7 Gyr, which suggests that the SFR may have been higher in the past. It is, however, more than two orders of magnitude lower than we might expect if the entire molecular ISM were forming stars,

$$\frac{\sum M_{\rm GMC}}{t_{\rm ff}} \simeq \frac{1.4 \times 10^9 \, M_\odot}{3.1 \, {\rm Myr}} \simeq 450 \, M_\odot \, {\rm yr}^{-1}, \tag{9.22}$$

where the free-fall timescale is based on an average cloud density $n_{H_2} \sim 10^8\,\text{m}^{-3}$ and the total molecular mass (tabulated in Chapter 10) includes helium.

This discrepancy between observed and predicted SFR was quickly realized once the link between molecular clouds and protostars was made in the 1970s. Evidently, despite their large non-thermal motions, GMCs are not in free-fall collapse. Magnetic fields may resist gravity on large scales in the diffuse regions of the cloud but their observed strengths are too low to support molecular clouds or the clumps and cores within.

We can express the efficiency of star formation in the form

$$\dot{M}_* = \epsilon_{\text{ff}} \frac{M_{\text{gas}}}{t_{\text{ff}}}, \tag{9.23}$$

where ϵ_{ff} is the fraction of the mass of a cloud that turns into stars on a free-fall time. Observations show that molecular clouds are generally inefficient star-forming factories with $\epsilon_{\text{ff}} \simeq 0.01$, varying by about a factor of 3 across all types of regions whether large or small and in different environments.

The measured cloud mass and radius vary with the observational technique. For instance, the CO and HCN observations of molecular clouds toward the Galactic Center in Figure 9.9 have critical densities that differ by more than two orders of magnitude (Table 7.1). In principle, we might expect ϵ_{ff} to depend on density and therefore to

Fig. 9.9. Molecular line emission toward the Galactic Center. The top panel shows CO emission from the bulk of the molecular gas with an average density $n_{H_2} \sim 10^8\,\text{m}^{-3}$. The bottom panel shows HCN emission from much denser gas with a density more than two orders of magnitude greater. The intensity scale varies from 0 to 90% of the map maximum in each case and demonstrates that the amount of dense, potentially star-forming, gas is a small fraction of the total amount of molecular material. The publicly available data are from the Atacama Submillimeter Telescope Experiment (ASTE) in Chile.

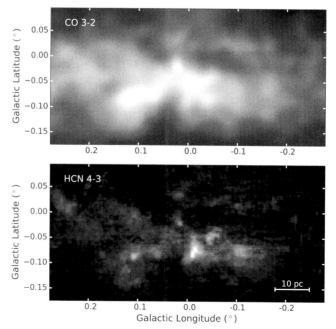

differ depending on the observational tracer. However, the small, dense cores seen in HCN only account for a small fraction of the total mass visible in CO and ϵ_{ff} remains approximately constant.

The empirical finding that stars form slowly, relative to the free-fall time, at all scales and densities appears to be due to the energy input from the young stars into their surroundings. Modeling this feedback loop is a major challenge and a topic of current research.

The overall **star formation efficiency** (SFE) is the mass fraction of a cloud that ultimately turns into stars,

$$\epsilon_* = \frac{M_*}{M_{gas} + M_*}. \tag{9.24}$$

This is a useful theoretical construct but hard to measure observationally since it is an integrated property over the star-forming history of a region and we only see a snapshot in time.

Most stars form in clusters and the efficiency of their formation dictates whether they remain bound or not as the non-star-forming gas is lost. We assume the cluster forms in a cloud with total mas $M_{tot} = M_{gas} + M_*$, and the gas and stellar distribution have the same radius R_0 and velocity dispersion σ. The total kinetic energy is $T = \frac{3}{2} M_{tot} \sigma^2$ and potential energy, $W = -\alpha G M_{tot}^2 / R_0$, where α is a factor of order unity that describes the radial density variation. From the virial theorem (Equation 8.29),

$$\sigma^2 = \alpha \frac{G M_{tot}}{3 R_0} = \alpha \frac{G M_*}{3 \epsilon_* R_0}. \tag{9.25}$$

If the gas is instantaneously removed such that the stars do not move, then the total energy of the cluster is

$$E = T + W = \frac{3}{2} M_* \sigma^2 - \alpha \frac{G M_*^2}{R_0}. \tag{9.26}$$

As the cluster dynamically evolves, it will expand to reach virial equilibrium at a radius R_1. Since the energy is constant,

$$E = \frac{1}{2}(2T + W) + \frac{1}{2} W = \frac{1}{2} W = -\alpha \frac{G M_*^2}{2 R_1}. \tag{9.27}$$

Combining these equations leads to a relation between the initial and final radius of the cluster,

$$\frac{R_1}{R_0} = \frac{\epsilon_*}{2\epsilon_* - 1}. \tag{9.28}$$

The ratio is unity for $\epsilon_* = 1$ and increases as ϵ_* decreases below that, corresponding to a greater fraction of the gravitational binding energy being lost. The final radius grows without limit as ϵ_* approaches 0.5, corresponding to an unbound cluster, and there is no physical solution for $\epsilon_* < 0.5$. Thus, clusters can only remain bound in an HII region or other situations where the gas is lost on dynamically short timescales if the overall SFE is greater than 50%.

If, however, the gas is lost more slowly such that the cluster remains in virial equilibrium during the process, then we can use the same reasoning as above but consider small steps, $R_1 = R_0 + dR$, $M_{\text{tot}} = M_{\text{tot}} + dM_{\text{tot}}$, to derive the simple differential form

$$\frac{dR}{R_0} = -\frac{dM}{M_{\text{tot}}}. \qquad (9.29)$$

Integrating with the same initial and final limits as for the fast case gives the ratio

$$\frac{R_1}{R_0} = \frac{1}{\epsilon_*}. \qquad (9.30)$$

In this case, clusters expand adiabatically and can remain bound, though they will become so large if the SFE is low that other agents will break them up. The two cases are illustrated in Figure 9.10.

The fraction of pre-main sequence stars in clusters is much greater than field stars in clusters. This implies that most clusters are unbound and formed with a low overall efficiency or a rapid loss of gas. Large clusters are likely to contain an OB star which can ionize and thereby remove much of the gas. Smaller groups of low-mass stars can also remove gas through protostellar outflows which we describe below. Since ϵ_{ff} is approximately constant, however, bound clusters do not form stars faster but are apparently able to continue building up their stellar mass over a greater number of free-fall timescales than their unbound counterparts. The takeaway is that feedback is an essential component of star formation.

Fig. 9.10. The change in cluster radius as gas disperses from a star-forming region as a function of the percentage star formation efficiency, ϵ_*. If the gas is removed faster than dynamical timescales, the cluster will become unbound for $\epsilon_* \leq 0.5$. More gradual gas dispersal allows the cluster to adjust and remain bound in all cases, though its final size can be very large for low ϵ_*.

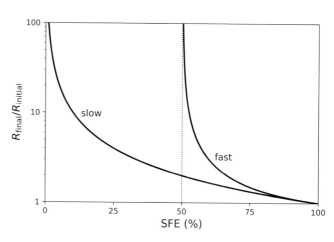

9.6 Angular Momentum

As noted in the preface to this chapter, the gravitational collapse of a cloud core to a star is accompanied by an enormous change in size scale, about seven orders of magnitude. In the same way as ice skaters spin faster as they pull in their arms, if all the angular momentum of the core were to be transferred to the star, it would rotate very rapidly.

The moment of inertia for a uniform sphere of mass M, radius R, is $I = \frac{2}{5}MR^2$. The prefactor changes slightly for different density profiles but the MR^2 scaling remains the same. The angular momentum is $J = I\Omega$, where Ω is the angular speed. Therefore, with no change in mass or angular momentum, we have

$$\Omega_{star} = \left(\frac{R_{core}}{R_{star}}\right)^2 \Omega_{core} \simeq 10^{13}\Omega_{core}, \tag{9.31}$$

for $R_{core} = 0.2$ pc, $R_{star} = 3R_\odot = 2 \times 10^9$ m. A lower limit to the angular speed of a molecular cloud and its substructures is that of the Galaxy. The Sun's orbital radius and speed are about 8 kpc and 230 km s^{-1} respectively, which imply a period of 220 Myr and rotation rate $\Omega \simeq 10^{-15}$ radian s^{-1}. This then places an upper limit to the rotation period of the star at an unphysically short $2\pi/\Omega_{star} \lesssim 10$ minutes. Stars cannot spin that fast because they would tear apart. Instead, the core may fragment as its density increases (see discussion in Chapter 8) and form more than one star. The bulk of the core rotation then goes into orbital motion rather than stellar rotation. Indeed, about 50% of field stars in the Galaxy are in binary or multiple systems. Magnetic fields are also very effective in coupling small and large scales and torquing down the stellar spin.

Ultimately, the part of a core that collapses to a single object will contract until gravity is balanced by centrifugal force. This only acts perpendicular to the rotational axis so the spherical core flattens to a disk. If we consider a test particle with mass m at radius R rotating at speed v, the ratio of the two forces is

$$\frac{F_{grav}}{F_{cen}} = \frac{GMm/R^2}{mv^2/R} = \frac{GM}{R^3\Omega^2}, \tag{9.32}$$

where $\Omega = v/R$ is the angular speed. Assuming no loss of angular momentum, $R^2\Omega = R_{core}^2\Omega_{core}$, and setting the force ratio to unity then gives the disk size scale,

$$R_{disk} = \frac{R_{core}^4\Omega_{core}^2}{GM_{star}}. \tag{9.33}$$

For the fiducial parameters we have used so far, this comes to 100 au but is clearly very sensitive to the initial core radius and angular speed.

Balancing forces does not necessarily imply a stable solution. Rather, continued infall from the core and the high disk surface density produce a very dynamical situation. Numerical simulations show that gravitational instabilities may form and shear into spiral waves, similar to disk galaxies. Fragmentation in the spirals provides a second avenue for binary star formation or for mutual interactions that transfer angular momentum outwards and mass inwards, thus growing the star.

The disk is rotationally connected to the protostar through strong magnetic fields (which are also concentrated during the collapse). Their mutual interaction produces an outflow that can carry away angular momentum. These are high-velocity, unbound, flows of gas that originate at the star–disk interface or within the inner few au of the disk. Observations show highly collimated, fast jets from very young, highly embedded protostars (Figure 9.11) and broader flows from later evolutionary classes. If the magnetic field is frozen into the gas, as the outflow expands radially it exerts a torque on the disk and slows it down. Outflows also entrain and sweep away the surrounding molecular core. This removes much of the surrounding mass reservoir and helps advance the evolutionary sequence along from embedded protostar

Fig. 9.11. Infrared image of the vibrational $v = 1 - 0$ line of H_2 at 2.12 μm produced by a protostellar outflow in Orion (named HH212 because of its prominence in this line) observed by the Very Large Telescope in Chile. The lower panel annotates the pertinent features, showing a jet from an embedded protostar and disk that is too deeply obscured to be visible in this image. The jet creates numerous bow shocks as it shocks the surrounding material. The outline of the remains of the dense core that formed the star is revealed through the scattered light nebulosity at the center. Image is courtesy of ESO.

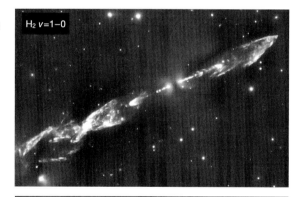

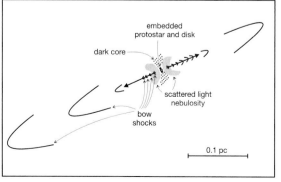

to optically visible pre-main sequence star. The combined effect of multiple outflows in a star-forming region injects enough energy and momentum to affect molecular cloud structure and dynamics.

Eventually, the mass infall rate decreases and the outflow stops. Young, rapidly rotating, pre-main sequence stars slow down as they lose angular momentum through stellar winds. Nevertheless, exoplanet searches show that, in most cases around low-mass stars, the spinning core leaves its final legacy in the form of orbiting Earths, Neptunes, and Jupiters. In the case of the Solar System, planets account for over 99% of its angular momentum but only 0.1% of its mass.

9.7 The Circumstellar Medium

Circumstellar disks have many similarities with the general topic of the ISM but also several fascinating differences. For this reason alone, they are worth the diversion here. In addition of course, our Solar System and the ubiquitous exoplanets around other stars formed in such disks.

Unlike the ISM, disks are not supported by thermal or turbulent pressure but by rotation. For a test particle of mass m at radius r around a star of mass M_*, the gravitational force is

$$F_g = \frac{GM_*m}{r^2} = \frac{mv^2}{r}, \tag{9.34}$$

where v is the rotational speed and v^2/r is the centrifugal acceleration. This gives the characteristic Keplerian rotational profile,

$$v_{\text{Kep}} = \left(\frac{GM_*}{r}\right)^{1/2} = 3.0 \left(\frac{M_*}{1\,M_\odot}\right)^{1/2} \left(\frac{r}{100\,\text{au}}\right)^{-1/2} \text{km s}^{-1}. \tag{9.35}$$

Even at the outer edges of a disk, rotational speeds are much greater than thermal speeds in cores and similar to turbulent motions in clouds. Speeds increase closer to the star and the non-constant angular rotation rate, $\Omega = v_{\text{Kep}}/r = (GM_*/r^3)^{1/2}$, produces a radial shear that has important consequences.

A particle of mass m at radius r has angular momentum $J = I\Omega = mr^2\Omega = (GM_*r)^{1/2}m$, and energy $E = T + W = mv^2/2 - GM_*m/r = -GM_*m/2r$. Now consider two particles, with the same mass at different radii, r_1, r_2, and assume without loss of generality that $r_1 < r_2$. Then their summed angular momentum and energy are

$$J = (GM_*)^{1/2}m \left(r_1^{1/2} + r_2^{1/2}\right),$$
$$E = -\frac{GM_*m}{2}\left(\frac{1}{r_1} + \frac{1}{r_2}\right). \tag{9.36}$$

Because the two particles rotate at different speeds, any friction in the disk will produce heat that will radiate away and cause the overall energy to decrease. The angular momentum, however, must stay the same as there is no external torque on the system. This implies changes in radii, $\Delta r_1, \Delta r_2$, which are related by

$$\Delta J = -\frac{(GM_*)^{1/2}m}{2}\left(\frac{\Delta r_1}{r_1^{1/2}} + \frac{\Delta r_2}{r_2^{1/2}}\right) = 0. \qquad (9.37)$$

The change in energy is therefore

$$\Delta E = \frac{GM_*m}{2}\left(\frac{\Delta r_1}{r_1^2} + \frac{\Delta r_2}{r_2^2}\right) = \frac{GM_*m}{2r_1^2}\left[1 - \left(\frac{r_1}{r_2}\right)^{3/2}\right]\Delta r_1. \qquad (9.38)$$

Since $r_1 < r_2$, $\Delta E < 0$ only if $\Delta r_1 < 0$ and therefore $\Delta r_2 > 0$. That is, the particles move apart from each other and we conclude that the disk evolves to a lower energy state by spreading out.

Although the total angular momentum is unchanged, it is redistributed. Because $J \propto r^{1/2}$ and friction makes particles move inwards, they lose angular momentum (and particles at larger radii move further out and gain angular momentum). Eventually the inner particles fall, or accrete, onto the star. This reasoning gives rise to the concept of an **accretion disk**. The mass accretion rate depends on the stress between two shearing layers which is equal to the velocity gradient times the **viscosity**.

Observations of pre-main sequence stars after the surrounding core has dispersed show (approximately blackbody) stellar photospheres with excess emission in the optical and ultraviolet. This is due to shocked gas accreting from the disk onto the star and the inferred mass accretion rates can be derived from the luminosity following Equation 9.15, with values $\dot{M} \sim 10^{-8}\,M_\odot\,\text{yr}^{-1}$. These rates are far too high to be explained by the viscosity from molecular diffusion that we are familar with in terrestrial fluids and implies that a different, likely non-ideal MHD process transports angular momentum in the disk. The accretion rate is observed to decrease with time but remains detectable in pre-main sequence stars up to ages $t \sim$ few Myr. This shows that disks are much longer lived than the cores from which they formed, and that their initial masses are at least $\dot{M}t \sim 10^{-2}\,M_\odot \sim 10\,M_{\text{Jupiter}}$, which is sufficient to form planetary systems.

The disks are warmer closer to the star than further away due to the heating by viscosity and stellar radiation. We can estimate the temperature scaling for the former by balancing the rate at which energy is lost as mass flows from r to $r + \Delta r$ (where $\Delta r < 0$) with luminosity

$$-\Delta E = -\frac{GM_*\dot{M}}{2r^2}\Delta r = 2A\sigma T_{\text{acc}}^4. \qquad (9.39)$$

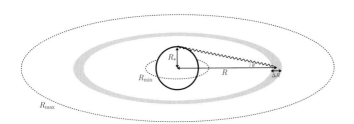

Fig. 9.12. Radiative heating of a flat disk by starlight. The size of the star is greatly exaggerated to illustrate the incident angle of radiation.

The last step assumes blackbody emission at temperature T_{acc} over annular area $A = 2\pi r \Delta r$ with an additional factor of 2 to account for both sides of the disk. Thus the temperature varies with radius as

$$T_{\mathrm{acc}} = \left(\frac{GM_*\dot{M}}{8\pi\sigma r^3}\right)^{1/4}. \tag{9.40}$$

This is a simplification that treats each annulus separately. A full treatment includes the work done by viscous torques from the inner disk. This changes the normalization but not the radial dependence.

For the stellar heating, we equate the radiation flux through an area A on the disk with its loss,

$$\frac{L_*}{4\pi r^2} A\theta = A\sigma T_{\mathrm{rad}}^4, \tag{9.41}$$

where θ is the angle between the path of the stellar radiation and the plane of the disk surface. Figure 9.12 illustrates the geometry and shows that $\theta \sim R_*/r$, from which we determine

$$T_{\mathrm{rad}} = \left(\frac{L_* R_*}{4\pi\sigma r^3}\right)^{1/4}. \tag{9.42}$$

This has the same $r^{-3/4}$ dependence as heating due to viscous accretion. For canonical values, $M_* = M_\odot, L_* = L_\odot, R_* = R_\odot, \dot{M} = 10^{-8}\, M_\odot\, \mathrm{yr}^{-1}$, the temperatures at 1 au are similar, $T_{\mathrm{acc}} = 65\,\mathrm{K}, T_{\mathrm{rad}} = 103\,\mathrm{K}$. Younger disks tend to have higher accretion rates and therefore higher temperatures, whereas stellar radiation dominates the heating in older disks as the accretion rate decreases.

With a given temperature profile, we can determine the thermal emission from a circumstellar disk. We assume that the disk is inclined with angle i (where $i = 0$ corresponds to face-on) and at distance d. The contribution from each region of the disk is the specific intensity times the solid angle,

$$dF_\nu = B_\nu(1 - e^{-\tau_\nu})\frac{dA}{d^2}, \tag{9.43}$$

where the optical depth at frequency ν is the product of the mass absorption coefficient and surface density, $\tau_\nu = \kappa_\nu^{\mathrm{dust}}\Sigma/\cos i$. Here Σ

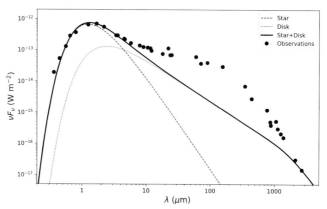

Fig. 9.13. The SED of a flat disk compared to observations of AA Tau, a pre-main sequence star in the Taurus star-forming region at a distance of 145 pc. The stellar contribution is modeled as a 3000 K blackbody shown by the dashed line. A flat disk profile is shown for an inclination of 60°, temperature of 80 K at 1 au and radial profile of $r^{-3/4}$, surface density of 10^3 kg m^{-3} at 1 au and r^{-1} profile. The integral is calculated over radii 0.01 au to 100 au. The disk emission is shown by the dotted line and the sum of the star and disk is the solid line. The circles are the observed values from optical to millimeter wavelengths. The discrepancy in the mid- and far-infrared implies that the disk is warmer at large radii than the flat disk model predicts.

is the surface density normal to the disk at each radius and the $\cos i$ accounts for the projection along our line of sight. Integrating over annuli leads to an expression for the total flux density,

$$F_\nu = \frac{2\pi \cos i}{d^2} \int_{R_{\min}}^{R_{\max}} B_\nu(T)(1 - e^{-\kappa_\nu^{\mathrm{dust}}\Sigma/\cos i})r\,dr. \qquad (9.44)$$

With a prescription for temperature as given above, we can then model the SED of a star–disk system. The surface density profile and radial range must also be specified. The effects of these parameters are a good student exercise to build intuition. An example is shown in Figure 9.13 and compared to observations of the flux density at multiple wavelengths of a pre-main sequence star.

Equation 9.44 shows that the disk SED is the sum of graybody profiles with different temperatures. This is why the emission extends over several decades in wavelength, primarily in the infrared, and falls off in the millimeter regime as the optical depth decreases. The flat disk model is able to explain the excess near-infrared emission found around pre-main sequence stars and the parameters can be adjusted to fit the millimeter wavelength data too. However, observations clearly show much stronger emission than predicted at wavelengths ~ 10–300 μm. Because different wavelengths map to different radii through the

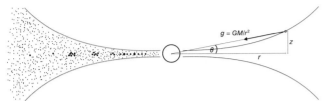

Fig. 9.14. Schematic of a disk, viewed edge on. The right hand side shows the gravitational force on a particle at radius R from the star and height z above the midplane. The vertical component of the force is proportional to the aspect ratio, z/R. The left hand side shows the dust particles within the flared disk. Small particles are swept along with the gas and extend over the full disk but the larger particles settle to the midplane and radially drift inwards. There they may agglomerate to become planetesimals and ultimately planets.

temperature profile, this excess in the mid- and far-infrared implies that the disk is warmer than expected beyond a few au from the star. After the first survey of the infrared sky in the 1980s and the characterization of hundreds of disks, it was quickly realized that the flat geometry required modification to allow the disk to intercept more starlight at large radii.

This leads us to consider the vertical structure of the gas. If we imagine a particle at radius r and height z above the disk midplane (right side of Figure 9.14), we see that the gravitational acceleration, with magnitude $g = GM_*/r^2$, can be split into radial and vertical components, $g_r = -g\cos\theta, g_v = -g\sin\theta$, where $\tan\theta = z/r$.

For small values, $z \ll r$, $\cos\theta \simeq 1$ so balancing the radial component against the centrifugal acceleration gives the Keplerian rotation rate described above. The gas is supported in the vertical direction by gas pressure through the equation of hydrostatic equilibrium,

$$\frac{dP}{dz} = -\rho g_v \simeq \rho \frac{GM_*}{r^2} \frac{z}{r}. \tag{9.45}$$

For an isothermal equation of state, $P = \rho c^2$, where c is the thermal sound speed. This results in a simple differential equation,

$$\frac{1}{\rho} \frac{d\rho}{dz} = \frac{\Omega^2}{c^2} z \equiv \frac{z}{h^2}, \tag{9.46}$$

where $h = c/\Omega$ is the scale height and the vertical density distribution varies as a Gaussian,

$$\rho(z) = \rho(0) e^{-z^2/2h^2}. \tag{9.47}$$

Moreover, because the gravitational force is weaker further from the star, the disk "atmosphere" becomes more extended at larger radii. The sound speed, $c \propto T^{1/2} \propto r^{-3/8}$, and the angular speed, $\Omega \propto r^{-3/2}$, together imply that $h \propto r^{9/8}$. The angle from the star to the disk surface

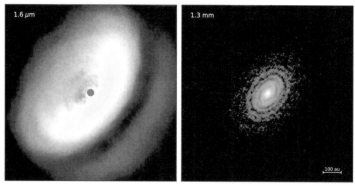

Fig. 9.15. The circumstellar disk around the pre-main sequence star IM Lup. The left panel, taken with the SPHERE instrument on the VLT, shows scattered light from small dust grains in the near-infrared. A coronagraph blocks out the light from the central star. The right panel, taken with the ALMA interferometer, shows thermal emission from large dust grains at millimeter wavelengths. Both images are $6'' \times 6''$ across. Credits: ESO/H. Avenhaus et al./DARTT-S collaboration and NRAO/DSHARP project.

increases gently as $\theta \simeq h/r \propto r^{1/8}$, which gives the flared structure that intercepts more stellar radiation than a flat disk.

A flared disk model can much better reproduce observed disk SEDs. The vertical structure and inner radius can then be determined, but other parameters, particularly density and outer radius, are more degenerate. However, a picture tells a thousand words and, through advances in technology, we can now spatially resolve disks and measure sizes, profiles, and map substructures. At optical and infrared wavelengths, a star can be very bright but, when masked by a coronagraph, the scattered light from small dust grains in the disk can be seen.

A spectacular example is shown in the left panel of Figure 9.15. The bright emission comes from the near-facing side of the disk and is strongly forward scattered. The opposite side has more contrast and shows hints of rings. The fainter band on the right side of the image shows light leaking through from from the back side of the disk. The dark belt between the two sides is the dense, dusty midplane that blocks out all starlight.

The right panel shows the same disk at millimeter wavelengths where the star is now very faint and no coronagraph is needed. This image was produced by the ALMA interferometer with multiple antennas on baselines that extend to 13 km, resulting in an extremely high angular resolution of $0.''04$. Following the discussion in Chapter 4, the emission reveals the location of dust grains that are approximately three orders of magnitude greater in size than those seen in the optical and near-infrared scattered light images; millimeters compared to microns. The striking difference between the two panels reveals several important physical principles. The millimeter emission shows a spiral pattern and is much more compact both radially and vertically than the near-infrared.

A massive planet may perturb the disk enough to produce spiral arms but, in this case, the two-armed feature is most likely caused by gravitational instability of the disk itself. The Jeans mass argument shows that pressure support resists gravity on small scales. The shear in a differentially rotating disk provides an additional means of support but now on large scales. As a back-of-the-envelope estimate, consider a cylindrical region of the disk at radius r, with radial extent Δr and extending perpendicular to, and above and below, the midplane. This will shear apart on a timescale $t_{shear} \sim \Delta r / [v(r) - v(r + \Delta r)] \sim 1/|dv/dr|$, which is equal to $2/\Omega$ for a Keplerian profile. The disk will be stable if the shear timescale is shorter than the free-fall timescale, $t_{ff} \sim 1/(G\rho)^{1/2}$. Expressing the volume density, $\rho \simeq \Sigma/2h$, the condition for gravitational stability is $c\Omega/2G\Sigma > 1$. A proper analysis gives the **Toomre criterion for stability**,

$$Q = \frac{c\kappa}{\pi G \Sigma} > 1, \tag{9.48}$$

where the epicyclic frequency, κ, refers to the oscillations of radial perturbations to an orbit (for a Keplerian disk, $\kappa = \Omega$). IM Lup is a cold, dense disk and the Toomre Q drops below unity at intermediate, $\sim$30 au, scales. This is not commonly seen in disks around optically visible pre-main sequence stars with ages of 1 Myr and greater, but more embedded disks are more massive and gravitational instability may be an important facet of their early evolution.

The flatter, smaller appearance of the millimeter disk shows that larger dust grains are more concentrated both vertically and radially (and thus the flat disk model in Equation 9.44 is a useful approximation at millimeter wavelengths). This is a distinguishing characteristic of disks compared to the general ISM and is a critical first step along the pathway to planets. The underlying reason is that the dust grains have a much lower velocity dispersion than the gas particles and do not respond as strongly to pressure gradients. When we consider the fluid equation for momentum conservation (Equation 8.5), the rotation rate of the gas satisfies

$$\frac{v_{rot}^2}{r} = \frac{GM_*}{r^2} + \frac{1}{\rho}\frac{dP}{dr}, \tag{9.49}$$

which can be restated as

$$v_{rot}^2 = v_{Kep}^2 + \frac{r}{\rho}\frac{dP}{dr}. \tag{9.50}$$

Because the gas is denser and hotter closer to the star, $dP/dr < 0$, and the gas rotates slower than Keplerian. Assuming thermal pressure $P = \rho kT/m$, and power law profiles $\rho \propto r^{-a}, T \propto r^{-b}$, the derivative term simplifies to $-(a + b)kT/m$. The fractional reduction in the rotation rate is then $(1 - \eta)^{1/2}$ where

$$\eta = (a + b)\left(\frac{c}{v_{\text{Kep}}}\right)^2. \tag{9.51}$$

The power law indices typically sum to about 2 and rotation rates are fast so η is small. At 30 au, $\eta \simeq 0.02$, so the gas rotation speed is about 99% of Keplerian. This seems like a small difference but the effect is dramatic for dust particles that are not supported by the pressure gradient and try to rotate at the Keplerian speed. They feel a headwind of $\sim 0.01 v_{\text{Kep}} \simeq 60 \, \text{m s}^{-1}$, which would be a Category 3 hurricane here on Earth.

Clearly then, we need to consider the aerodynamic effects on the dust. We first note that the gas density at 30 au scales in disks like IM Lup is of order $n_{\text{H}_2} \sim 10^{12} \, \text{m}^{-3}$. This implies a mean free path, $l = 1/n_{\text{H}_2}\sigma_{\text{H}_2} \simeq 10^{-4}$ au, so we need to consider individual interactions between the dust and gas particles. A dust grain with mass m and size a moving at speed v relative to the gas will be slowed down by collisions. Over a time t the particle sweeps through a mass of gas $m_{\text{gas}} = \pi a^2 \rho v t$. The stopping time is defined to be when the momentum of the displaced gas, $m_{\text{gas}}c$, equals the momentum of the particle, mv. Because grain sizes vary more than their density, this is often expressed in the following way,

$$t_s = \frac{4}{3}\frac{a}{c}\frac{\rho_{\text{grain}}}{\rho}, \tag{9.52}$$

where $\rho_{\text{grain}} = m/\frac{4}{3}\pi a^3$ is the average density of an individual grain (not the collective dust mass density in the disk). Gas drag is dynamically important if t_s is small compared to the rotation timescale. This condition is expressed through the dimensionless **Stokes number**,

$$\text{St} = t_s\Omega. \tag{9.53}$$

If $\text{St} < 1$, the stopping time is shorter than the orbital time and the dust–gas interactions are strong. Alternatively, $\text{St} > 1$ implies that the particles feel little effect from the gas headwind. For any given location in the disk, the value of the Stokes number scales linearly with the grain size. At 30 au, $\text{St} \simeq 1$ for $a \simeq 1 \, \mu\text{m}$. Smaller grains are coupled to the gas and larger grains are decoupled (for sizes larger than the mean free path, we have to consider a different drag prescription but the conclusion holds). Intermediate sizes feel the headwind and are small enough that they lose significant angular momentum and drift inwards. A useful terrestrial analogy can be seen by looking at clouds in the sky: when the water droplets are small they blow around with the wind as a white fluffy cloud, but as the droplets agglomerate they eventually become too heavy to be supported by gas pressure and fall down as rain.

In much the same manner, pressure supports the gas perpendicular to the disk but dust grains with St > 1 settle to the disk midplane. In this case, a terrestrial example is the sedimentation of silt in a stream. The vertical settling and radial drift is illustrated on the left hand side of Figure 9.14. The millimeter-emitting grains lie in a very thin layer at the midplane and at relatively small radii compared to the gas and micron-sized grains. The high dust concentrations promote further grain growth and, once the collective dust density becomes greater than the gas density, the dust drags the gas rather than the other way round, resulting in new instabilities and the formation of planetesimals. These are witnessed in the appearance of primordial asteroids and comets as "rubble piles". Fundamentally, the aerodynamic separation of solids from gas is the reason why disks form planets and cores form stars.

Notes

Star formation is a thriving research field. ALMA and other observational facilities are broadening our perspective to planet-forming disks around nearby stars and molecular clouds in the most distant galaxies. There are many textbooks on this subject, varying from introductory (Ward-Thompson and Whitworth, 2015) to detailed and comprehensive (Krumholz, 2017), and one focused entirely on planet formation (Armitage, 2020). For up to date reviews of current research, see the review by McKee and Ostriker (2007) and the Buether et al. (2014) conference book. The reference for the meta analysis of the Galactic star formation rate and stellar mass is Licquia and Newman (2015). Access to the data and papers on the ground-breaking high-resolution ALMA images of circumstellar disks are available at bulk.cv.nrao.edu/almadata/lp/DSHARP.

Questions

1a. Numerically integrate the Emden–Chandrasekhar Equation 9.8 by integrating outwards in small steps, Δx, using Taylor expansions,

$$y''(x) = e^{-y(x)} - 2y'(x)/x,$$

$$y(x + \Delta x) = y(x) + y'(x)\Delta x + y''(x)\Delta x^2/2,$$

$$y'(x + \Delta x) = y'(x) + y''(x)\Delta x,$$

where $y' = dy/dx, y'' = d^2y/dx^2$. By symmetry, the boundary conditions are $\phi = 0, d\phi/dr = 0$ at $r = 0$ which translate to $y =$

$y' = 0$ at $x = 0$. (The limiting value of $y''(0) = 1/3$ can then be derived but the solution is not sensitive to this value.)

Create a series of density profiles for different central densities as in Figure 9.1.

1b. Calculate the mass of the core using Equation 9.11 and recreate Figure 9.2.

2a. Model the spectrum of a collapsing core using Equations 9.12 and 9.13. Plot spectra for the CS $J = 2-1$ line with different combinations of parameters including those in Figure 9.5.

2b. Suppose you were to observe the same transition of the isotopologue $C^{34}S$ but the optical depths τ_{f0}, τ_{b0} are lower by a factor of 10 due to its lower abundance. What does the line look like for the same parameters? (This sort of observation is carried out to test whether self-absorption creates the double-peaked profile in a core or whether there are two unrelated kinematic components along the line of sight.)

3a. Assume a pure power law IMF, $dN/d \ln M \propto M^{-\Gamma}$ for $M > 0.5 \, M_\odot$ and a Salpeter slope, $\Gamma = 1.35$. The number of stars in any given bin is an integer quantity, with Poisson statistics. If you observe a cluster with 1000 stars above $0.5 \, M_\odot$, how many O stars, $M > 16 \, M_\odot$, do you expect to find?

3b. How big would a (young) cluster be for it to have a 50% chance of containing at least one very massive star, $M > 50 \, M_\odot$?

Chapter 10
The ISM on the Galactic Scale

Up until now, we have discussed the various phases of the ISM mostly in isolation from each other and on relatively small spatial and temporal scales. Here we consider the relation between the different atomic, molecular, and ionized gas components across the Galaxy and over gigayear timescales. From this perspective, we see a dynamic ISM that acts as a recycling plant from which stars are born and to which they return. The associated gain of metals and loss of gas drives the long-term evolution of the Galaxy.

10.1 Distance Determination

A first step in learning about the large-scale distribution of the ISM is to measure distances to the sources that we observe. For the gas, we can use the Galactic rotation curve to invert the velocities derived from spectral line observations. For the dust, we can compare stellar reddening with measures of their parallax. We describe each method in turn.

Assuming a flat disk and circular orbits with speed V at Galactocentric radius R, we can use simple geometrical considerations to derive the radial velocity V_{rad}, of an object at longitude l, relative to the Sun,

$$V_{rad} = V(R) \cos \theta - V_S \sin l$$
$$= \left[V(R) \left(\frac{R_S}{R} \right) - V_S \right] \sin l, \qquad (10.1)$$

where the angle θ between the source velocity vector and line of sight is defined in Figure 10.1, $R_S = 8.2 \, \text{kpc}$ is the distance of the Sun from the Galactic Center, and $V_S = 220 \, \text{km s}^{-1}$ is the orbital speed of the Sun.

A given line of sight, at fixed l, intercepts a range of radii. At the tangent point where $\theta = 0$, the radius $R_{tang} = R_S \sin l$, and radial velocity

Fig. 10.1. The geometry of the rotation speeds of the Sun and an object at radius R and longitude l. The sightline from the Sun intercepts the inner orbit at two points, A and B. The observed radial velocity, V_{rad}, is the projection of the rotation curve, $V(R)$, by an angle θ. GC labels the Galactic Center.

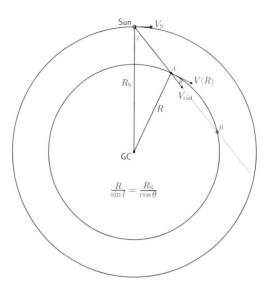

$V_{rad} = V(R_{tang})$. Because this is the maximum observed value for a gently varying $V(R)$, we can use HI or CO observations along different lines of sight to assign the highest velocity measured in each spectrum to the geometrically determined R_{tang} and thereby define the rotation curve. We can supplement this method and extend to the outer galaxy, $R > R_S$, through observations of HII regions. These are bright enough to be detected at large heliocentric distances with radial velocities readily measured from gas spectral lines and distances determined by comparing OB stellar brightness to the ionization rate.

The radial speed increases approximately proportional to radius (solid body rotation) in the inner 5 kpc and is a constant $220\,\text{km s}^{-1}$ at larger radii. Flat rotation curves are also measured in other spiral galaxies, where our perspective allows $V(R)$ to be more directly determined, and are characteristic of massive dark matter halos.

With the rotation curve established, we can invert Equation 10.1 to determine the radius for any velocity, not just the maximum in the spectrum. However, our vantage point within the disk results in blending of multiple features along the line of sight. For instance, all points toward or away from the Galactic Center have zero radial velocity, so their distances cannot be determined using this method. There is also a distance ambiguity in the inner galaxy, in the sense that points A and B in Figure 10.1 have the same V_{rad}. Nevertheless, the **kinematic distance** method is a simple and powerful technique. Non-circular motions limit its accuracy to $\sim 10-20\%$, but it is especially useful in the outer Galaxy where there is no distance ambiguity and where there are few stars to provide alternative indicators.

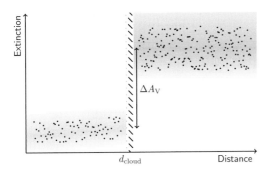

Fig. 10.2. Schematic of the dust screen model for determining the distance to a cloud based on a plot of stellar extinction versus distance. The former is determined from optical–infrared estimates of the reddening, the latter from Gaia parallax and modeling of stellar properties. The points represent individual stars and the gray scaling illustrates the dispersion, which is generally much less than the extinction jump due to an intervening cloud.

In recent years, the Gaia satellite has measured stellar parallaxes to a few tens of micro-arcseconds for hundreds of millions of stars. Combining the inferred distances with the reddening and extinction determined from all-sky optical and infrared surveys provides an important new way to measure cloud distances. Figure 10.2 shows that the extinction is initially low, then sharply rises by an amount ΔA_V related to the cloud column density, and then flattens off again. Modeling the cloud as a dust screen allows its distance, and in some cases its depth and structure, to be determined to an accuracy of about 5%. Because the method is optically based, it is limited to within approximately 2 kpc but complements the kinematic method, which performs better at larger radii.

10.2 The Distribution of Neutral Gas

With the location of different HI and CO features determined, we can convert angles to physical sizes, column densities to masses, and determine the distribution of the atomic and molecular gas as a function of Galactocentric radius, R, and height above the disk midplane, Z.

Up close, we find relatively little dust reddening and extinction in the solar vicinity. The Sun lies in the **Local Bubble**, a region spanning $60 - 100$ pc radially and $120-180$ pc vertically that is composed of diffuse ionized gas with a density $n_H \sim 10^5 \, \text{m}^{-3}$.

Beyond the Local Bubble, 21 cm observations find HI emission in every direction (see Figure 1.1) though the intensity can vary by an order of magnitude over a range of scales from arcminutes to degrees. HI is detectable out to $\sim$60 kpc from the Galactic Center, almost four times further than the stellar disk. Similarly large gaseous disks are also seen in other galaxies and provide important probes of galactic structure and dynamics far out into the dark matter halo. Between $R \sim 10-30$ kpc,

Fig. 10.3. Radial profiles of the surface density of atomic and molecular hydrogen gas across the Galaxy, in units of total number of hydrogen nuclei per square meter (scale on left hand axis). The dotted lines show the thickness of the HI and H_2 disks (full width at half maximum, scale on the right hand side).

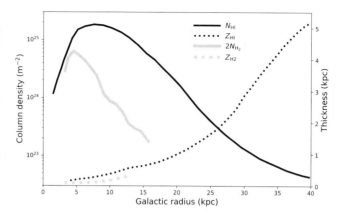

the inferred HI column density declines exponentially, $N_{HI} \propto e^{-R/R_{HI}}$, with scale length $R_{HI} = 3.75$ kpc. Within the inner Galaxy, as N_{HI} exceeds 2×10^{25} m^{-2}, corresponding to $A_V = 1$ mag, the hydrogen begins to turn molecular and the HI column density profile flattens (Figure 10.3).

Observations of CO trace the molecular gas, though with a notable gap between $A_V \sim 1-4$ mag, where "dark" H_2 is hard to detect (see PDRs in Chapter 7). To convert the observed CO integrated line flux, W_{CO}, to an H_2 column density adds additional uncertainty through the abundance, excitation, and optical depth. Generally, a simple linear conversion factor is assumed, $X_{CO} = N_{H_2}/W_{CO}$. This is discussed further when comparing the Galaxy to others in the following chapter. The inferred H_2 column density, shown as the dotted gray line in Figure 10.3, declines beyond ~ 4 kpc, with a similar exponential radial scale length as HI but a factor ~ 20 lower in absolute value.

The profiles fall off in the central region of the Galaxy. The peak lies approximately halfway between the Sun and Galactic Center and is known as the molecular ring. The structure within the ring is hard to determine but asymmetries in the stellar distribution suggest a bar. The gas then concentrates again within ~ 200 pc of the Galactic Center in the **central molecular zone**. The ISM in this region has distinct properties due to the influence of the supermassive black hole at the Galactic Center. We return to this in a broader context in the following chapter.

Turning to the vertical distribution of the gas, Figure 10.4 plots the HI column density across the Galaxy, stretched along the latitude axis to emphasize the deviation from the midplane. Unlike Figure 1.1, which used a non-linear scaling to show both faint and bright features, the grayscale here is directly proportional to column density. Most of the emission beyond the first tickmark at $l > 45°$ is from gas in the

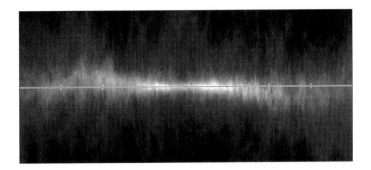

Fig. 10.4. The HI column density across the Galaxy, shown from Galactic Longitude $l = -180°$ to $+180°$ and Galactic Latitude $b = -20°$ to $+20°$ on a linear scale. The line shows the Galactic midplane $b = 0$ with ticks at 45° intervals. The dataset is from the HI4PI survey.

outer Galaxy. The column density becomes noticeably lower and the distribution departs from the midplane, to higher latitude at positive longitude and lower latitude at negative longitude (left and right sides of the figure respectively). This warp in the Galactic disk extends 5 kpc above and below an extrapolation of the inner Galaxy plane. Its origin is unknown but it is not an unusual feature in spiral galaxies and may be due to self-gravitational instabilities (related to the Toomre Q parameter in Equation 9.48) or perturbations from neighboring or accreting galaxies.

If we look at deviations from the warped midplane, we can also see that the scale height of HI increases with radius. The disk is flared due to the decreased gravitational force in the outer regions, analogous to circumstellar disks in Chapter 9 but with an extended dark matter halo rather than a central point source. The HI thickness, defined by where the column density drops to half its maximum value above and below the plane, is plotted as a dotted line in Figure 10.3 with scale on the right hand side. It is approximately exponential with a radial scale length of 10 kpc. There is little CO emission in the outer Galaxy but the molecular clouds that are detected follow the warp in the HI midplane. The thickness of the molecular layer is about half that of the HI and is only 100 pc in the inner Galaxy.

These distributions are azimuthally averaged but there are variations at a given radius due to spiral arms. The contrast in the HI brightness is generally small because of the ubiquity of emission, the wide linewidths in the warm gas, and distance confusion from non-circular motions associated with the streaming of gas through the arms. The CO maps show a much higher contrast, however, indicating that H_2 is more concentrated in spiral arms than HI.

The differences are demonstrated in the spectra plotted in Figure 10.5. Along this line of sight, through the inner regions of the Galaxy, negative radial velocities correspond to $R > R_S$. There is little CO emission here as molecular gas is sparse in the outer Galaxy. There is stronger emission in the inner galaxy, at positive velocity, but

Fig. 10.5. Spectrum of HI and CO along a line of sight through the inner Galaxy at $l = 35°, b = 0°$. Both spectra are shown in terms of brightness temperature and the CO scale is magnified by a factor of 10.

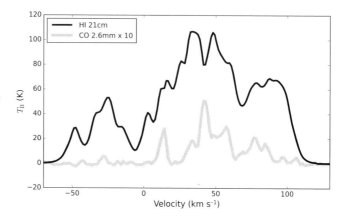

the broadness of the HI emission translates to a wide range of radii. The narrower features in the CO emission correspond to more radially concentrated regions of molecular gas. The CO peaks within a broad HI envelope but there is no strong detailed correlation and even an anti-correlation at the strongest CO peak where the HI dips. This is a signature of cold, absorbing gas associated with a molecular cloud, as shown in detail below.

Along lines of sight to distant radio continuum sources such as quasars, or background HI at the same projected velocity such as point A and B in Figure 10.1, 21 cm absorption measurements can tell us the amount and temperature of the ISM. By decomposing emission spectra such as those in Figure 10.5 into a series of Gaussian components, we can then recover not only the radial distribution of the atomic gas but also the proportion of cold and warm gas.

As noted in Chapter 5, some gas has intermediate temperatures of a few thousand kelvin and lies in the unstable regime between the classical CNM/WNM phases. This is likely due to the relatively long timescale required to achieve pressure balance compared to dynamical disturbances in the ISM such as from supernovae, HII regions, and protostellar outflows. Nevertheless, the CNM tends to lie closer to the midplane and does not extend as far out in radius as the WNM. This is as expected since the combined stellar weight (or more precisely gravitational force in the hydrostatic equation) is greater within the inner Galaxy and toward the midplane, increasing the ambient ISM pressure. This gives some reassurance that, despite the dynamical activity, the underlying considerations behind the concept of a two-phase atomic medium are broadly applicable.

The denser CNM also tends to be concentrated in clouds, or less regular structures, that are surrounded by warm gas. This hierarchical

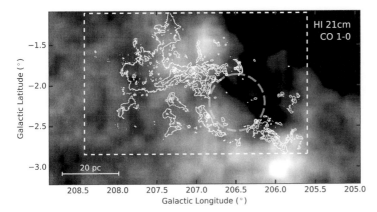

Fig. 10.6. Atomic and molecular gas in the Rosette molecular cloud. The grayscale shows 21 cm emission from the Arecibo radio telescope and the contours show the same CO $J = 1 - 0$ map as in Figure 7.7. The dashed white line shows the boundary of the CO map. The long dashed circle is centered on the HII region.

arrangement extends down to smaller, denser scales as we consider the molecular gas. When we look at HI toward our prototype GMC, the Rosette cloud, in Figure 10.6, we see that the emission is concentrated on the boundaries of the CO, especially around the HII region. The energetic radiation from this region ionizes and dissociates the surrounding gas.

The GMCs that lack OB stars are also surrounded by HI halos. This leads to a picture of gas condensing to molecular clouds out of a large atomic reservoir and then returning to a more diffuse atomic and ionized form as stars form. In this Galactic-scale context, molecular clouds represent a short-lived dense phase, occupying a tiny volume of the ISM, in the ongoing cycle of gas to stars and back again.

10.3 The Distribution of Ionized Gas

The bulk of the Galaxy, by volume, is filled with diffuse gas at densities $n_{\rm H} < 10^6 \, {\rm m}^{-3}$. Because ionization is faster than recombination at such low densities, most of this gas is ionized. With limited radiative pathways for cooling, gas temperatures range from warm, $\sim 10^4$ K, to hot, $\sim 10^6$ K.

Referring back to Chapter 6, the ionized phases of the ISM can be observed through bremsstrahlung continuum emission at long wavelengths and recombination lines from optical to radio. The strength of the emission in each case scales with the emission measure and therefore traces the relatively dense WIM rather than the very diffuse HIM.

Following the distribution of the WIM over Galactic scales is harder than for the neutral components discussed above. The radio continuum has strong synchrotron components and the strongest recombination line, Hα, lies in the optical and is affected by dust extinction. Pulsar

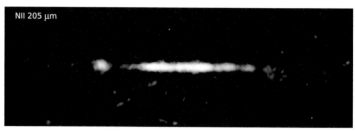

NII 205 μm

Fig. 10.7. The NII 205 μm line emission in the Galaxy. The map size is −180° to +180° in Galactic Longitude and −60° to +60° in Galactic Latitude (the same as the multi-wavelength panoramas in Figure 1.1). The brightness represents the line intensity and varies from 1.5 to 20 nW m^{-2} sr^{-1}. The data are from all-sky maps created by the Cosmic Microwave Background Explorer.

dispersion provides a way to determine column densities along numerous "pencil beam" lines of sight though generally limited by sensitivity to a few kpc from the Sun. Fine structure lines in the ground state of ionized nitrogen, NII, at 122 μm and 205 μm help fill out the picture. Nitrogen has an ionization potential of 14.5 eV which is only slightly higher than that of hydrogen, the low-energy levels are readily excited, and their emission in the far-infrared can be seen across the Galaxy, at least once the detector is above the Earth's atmosphere. An all-sky map of the 205 μm NII line is shown in Figure 10.7. A comparison with the Hα map in Figure 6.8 shows the limited sensitivity of this map but also its penetration along the Galactic plane. The emission is highly concentrated along the midplane and toward the inner Galaxy.

This old, noisy NII map provides a useful measure of the star formation rate. The observed line luminosity in the inner Galaxy, $L_{205} = 6.8 \times 10^6 \, L_\odot$, relates to the number of collisional excitations,

$$L_{205} = n_e n_{\mathrm{NII}} \gamma_{205} h\nu \, V_{\mathrm{NII}}, \qquad (10.2)$$

where $\gamma_{205} = 6.7 \times 10^{-14} \, \mathrm{m^3 \, s^{-1}}$ is the collisional rate coefficient, $h\nu = 9.7 \times 10^{-22} \, \mathrm{J}$ is the photon energy, and V_{NII} is the volume of the emitting region. The ionization rate equates with the number of recombinations and has a similar density dependence,

$$\dot{N}_{\mathrm{ionize}} = \alpha_2 n_e n_{\mathrm{HII}} \, V_{\mathrm{HII}}, \qquad (10.3)$$

where $\alpha_2 = 3.1 \times 10^{-19} \, \mathrm{m^3 \, s^{-1}}$ is the volumetric recombination rate at 8000 K. Assuming that the NII traces the HII, we can then deduce the total ionization rate in the inner Galaxy,

$$\dot{N}_{\mathrm{ionize}} = \frac{\alpha_2}{\gamma_{205}} \frac{L_{205}}{X_{\mathrm{NII}} h\nu} = 2 \times 10^{53} \, \mathrm{s^{-1}}, \qquad (10.4)$$

where $X_{NII} = [NII]/[HII] = [N]/[H] = 6.9 \times 10^{-5}$ is the nitrogen abundance relative to hydrogen and we implicitly assume that both elements are fully (singly) ionized throughout the regions of emission. This corresponds to about 10^4 O6 stars with typical ionizing luminosities $\sim 10^{49}$ s^{-1}. Averaged over a Salpeter IMF, the ionizing luminosity, $\langle \dot{N} \rangle \simeq 10^{46}$ s^{-1}, stellar mass, $\langle M_* \rangle = 0.5\,M_\odot$, and main sequence lifetime, $\langle t_{ms} \rangle = 3.7\,$Myr, imply a star formation rate, $\dot{M}_* = 2.7\,M_\odot\,$yr^{-1}, which is similar to other estimates though a little higher than the meta-analysis of all studies alluded to in Chapter 9.

The NII emission extends off the plane with an exponential scale height of 1 kpc. The star-forming molecular gas is far more concentrated toward the plane but the accompanying HII regions can break out and produce an extended low-density WIM across the entire Galaxy. However, observations toward pulsars with known distances show that the number density inferred from emission measures, $\propto n_e^2$, and pulsar dispersion, $\propto n_e$, can only be reconciled if the WIM is clumped along the line of sight with a filling factor $\phi \sim 20\%$.

The WIM is prevented from expanding to a more uniform density by the HIM. This is much harder to detect in emission or dispersion because the densities are orders of magnitude lower. It is best studied through UV absorption lines of highly ionized oxygen, OVI, or nitrogen, NV. Consequently, we do not have as precise a picture of its radial distribution, but it appears that the HIM occupies most of the volume of the ISM above about 400 pc above the midplane and extends many kpc into the Galactic Halo. Crucially, however, it remains bound to the Galaxy and can therefore return to the denser phases.

10.4 Galactic Recycling

The overall mass and filling fractions of the different phases of the ISM are summarized in Table 10.1. The filling fractions vary substantially with height above the midplane and are hard to determine precisely. The midplane is predominantly neutral whereas the halo mainly contains hot ionized gas that is so diffuse as to amount to a negligible mass component. The amount and distribution of these ISM components dictate where and how many stars form in the Galaxy and, through a complex feedback, the stars provide the energy that heats and stirs the gas, and in their death enrich the ISM with metals. The Galaxy holds the system together and slowly evolves on a characteristic timescale, $t \sim M_{ISM}/\dot{M}_* \sim 3\,$Gyr, that is much shorter than the Hubble time.

The structure of the ISM is far from smooth and different regions have been variously described as bubbles, shells, sheets, filaments, or tunnels and the overall appearance as turbulent, frothy, or fractal.

Table 10.1. Galactic distribution of ISM components

Phase	$M\ (M_\odot)$	$\phi_{Z<0.3\text{kpc}}$	$\phi_{Z>1\text{kpc}}$
Molecular	1.4×10^9	< 0.01	0
CNM	5.4×10^9	~ 0.1	0
WNM	5.4×10^9	~ 0.4	0
WIM	2.7×10^9	~ 0.3	~ 0.2
HIM	$\sim 10^8$	~ 0.2	~ 0.8

Although pressure may determine the general distribution of the atomic gas, the HIM is produced by supernovae shocks in an extremely dynamic ISM.

As discussed in Chapter 9, the stellar IMF declines steeply with mass, $\xi(M) = dN/dM_* \propto M_*^{-2.35}$ for $M_* > 1\,M_\odot$. This implies that most stars, whether by number or mass, are of lower mass than the Sun. These have an important, localized effect when they born, through protostellar outflows, and when they eject matter in the late stages of their life as red giants. However, the steep dependence of luminosity on mass, $L_* \propto M_*^{3.5}$, implies that the total luminosity of a collection of stars, $L_{\text{tot}} = \int L_* \xi dM \propto M_*^{2.15}$, is dominated by the upper end of the mass distribution. Moreover, only OB stars are hot enough to produce enough UV radiation to significantly ionize the ISM.

Because the rarest stars have the largest impact, stellar feedback on the ISM is an inherently stochastic process. The biggest effects come from the largest groups, not just because many OB stars produce more energy but also because they collectively act in the same region over a much longer time than any individual star. Over $\sim 10\,$Myr timescales, a cluster of OB stars can produce a **superbubble** that can reach hundreds of parsecs in size. The extent is so great that superbubbles break through the molecular and CNM disk into regions of lower pressure, such that they become more elongated vertically than radially.

Figure 10.8 demonstrates the immense scale of superbubbles as seen from the HI shells that surround them. The bubble on the far left is almost fully evacuated over a huge region about 600 pc across. The outer shell of swept-up atomic gas has a mass of $4 \times 10^6\,M_\odot$ and is expanding at a velocity of 22 km s^{-1}. For a power law increase in size with time, $R \propto t^p$, then the velocity scales as $v \propto pR/t$, and the dynamical age is $t = pR/v$. The expansion is similar to that of supernovae and HII regions considered in Chapter 8 but with an approximately constant energy input such that the kinetic energy of the expanding region grows linearly with time, $Mv^2 \propto t$. Assuming uniform density surroundings, this implies the radius grows as $R \propto t^{3/5}$ and we therefore infer an age

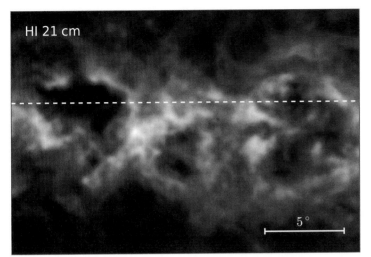

HI 21 cm

$5°$

Fig. 10.8. Superbubbles in atomic gas, as seen toward the region $l = 270°$ along the Galactic plane (the dotted line marks $b = 0°$). Note the angular scale bar compared to the Rosette molecular cloud image in Figure 10.6. The physical size depends on the distance to each object; the bubble on the far left is about 600 pc in diameter. The dataset is from the HI4PI survey.

for the bubble of $t \simeq 9$ Myr. The total kinetic energy in the expanding shell is $E = 2 \times 10^{45}$ J but numerical models that account for radiative losses show that the stellar input is about 10 times greater. Consequently the total energy requirement to produce the observed feature is $10E$ which corresponds to the radiation, winds, and supernova impact from ~ 100 OB stars.

As bubbles continue to expand, their appearance in HI may change from shell-like to open-topped structures with the various terminology of "fountains", "chimneys", or "worms". These offer low resistance to gas and ionizing radiation to stream up into the halo. What goes up must come down, at least for a massive galaxy like ours. Ultimately, pockets of the expanding gas cool, become denser and atomic, and fall back down to the plane. This creates some of the **high-velocity clouds** that are found at high Galactic latitude.

A schematic edge-on view of the Galactic ISM disk is shown in Figure 10.9. It shows the concentration of molecular clouds and CNM along the warped midplane, a wider disk of lower density WNM, the escape of ionizing radiation from HII regions in an inhomogeneous medium to produce the diffuse WIM, and the creation of the HIM in supernovae, their sustenance through mergers, escape to high latitude in superbubbles, and relatively cooler denser gas then falling back down as high-velocity clouds. This is only a conceptual guide, leaving out many details and open questions that are the subject of current research. In addition, magnetic fields and cosmic rays are important components in a complete description of the Galactic ISM that are not included here. Starting points to explore the full complexity and richness that more detailed observations and models provide are given in the Notes section.

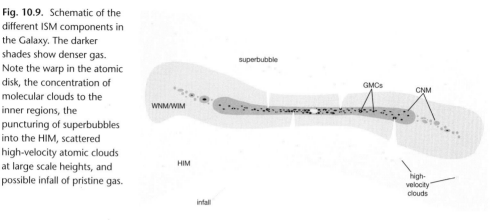

In tandem with its physical rearrangement, the gas is enriched in metals by supernovae and winds from the late stages of evolution in stars of all masses. This Galactic chemical evolution is a large subject in its own right but worth a short diversion here to explore one additional aspect of the ISM.

We define a system of gas and stars with masses M_{gas}, M_*, respectively. The mass of heavy elements is $M_Z = Z M_{\mathrm{gas}}$ where Z is the metallicity. All are functions of time t and initially $M_* = Z = 0$. The formation of a stellar mass, ΔM_*, locks up a mass of heavy elements, $\Delta M_Z = -Z \Delta M_*$, as most stars are low mass with long lifetimes. Conversely, the high-mass stars have very short lifetimes and enrich the gas by an amount $\Delta M_Z = y \Delta M_*$ where y is the yield. Ignoring enrichment from the end states of lower mass, longer lived stars that formed before, the overall change is

$$\Delta(Z M_{\mathrm{gas}}) = (y - Z)\Delta M_*. \tag{10.5}$$

If we further assume that the total mass, $M_{\mathrm{tot}} = M_{\mathrm{gas}} + M_*$, is constant, then eliminating $\Delta M_* = -\Delta M_{\mathrm{gas}}$ results in a simple differential eqation, $M_{\mathrm{gas}} \Delta Z = -y \Delta M_{\mathrm{gas}}$, with solution,

$$Z(t) = y \ln \left[\frac{M_{\mathrm{tot}}}{M_{\mathrm{gas}}(t)} \right]. \tag{10.6}$$

As the gas transforms into stars and M_{gas} decreases, the metallicity increases due to enrichment from the new massive stars that quickly die. The assumption of instantaneous recycling can be relaxed in more detailed models but the basic picture holds and ultimately the gas becomes highly enriched at late times. The present-day values for the Galaxy, $M_{\mathrm{tot}}/M_{\mathrm{gas}} = 10, Z = 0.02$, give a yield $y \simeq 0.01$.

The dynamic ISM quickly homogenizes the metals into the gas but the stars show a record of the enrichment, $Z(t)$. Here, the model breaks down as it predicts many more old stars with low Z than are observed. A proposed solution is to relax the closed box assumption that M_{tot} is constant and allow the accretion of gas into the system. In this scenario, the Galaxy built up from a low initial mass of gas, producing stars at a steady rate that continually enrich the system: both the gas and subsequent stellar generations. The lockup of metals into long-lived stars and yield from instantaneous recyling remains the same and Equation 10.5 still applies if the infalling gas is primordial ($Z = 0$). But the total mass is no longer constant so $\Delta M_* + \Delta M_{\text{gas}} = \Delta M_{\text{tot}} > 0$. Further steps require a prescription for the mass infall rate. The simplest mathematical solution is if the infall rate balances the star formation rate and the gas mass remains constant, $\Delta M_{\text{gas}} = 0$. This then results in the metallicity dependence,

$$Z(t) = y \left[1 - e^{[1 - M_{\text{tot}}(t)]/M_{\text{gas}}} \right]. \tag{10.7}$$

For the Galaxy, $M_{\text{tot}}/M_{\text{gas}} = 10$ and $y \simeq Z = 0.02$. This is a high yield but plausible. The implication for the ISM that results from the observed lack of low metallicity stars is that at least some of the gas that goes into stars is replenished by infall of primordial gas.

A grand synthesis of the ISM that incorporates all the dynamics, heating, cooling, star formation, stellar death, and chemical enrichment is not analytically tractable but amenable to numerical simulations that will become ever more detailed as computing power increases. To help guide such work, we can also look at the ISM in different galactic environments.

Notes

The stellar extinction method for determining distances to clouds has more subtleties than described here since it requires a priori assumptions about the stellar temperatures and needs to account for observational biases. More details can be found in Zucker et al. (2019). Figure 10.3 is based on HI and H_2 profiles presented in Kalberla and Kerp (2009) and Heyer and Dame (2015) respectively. These two reviews provide good starting points for exploring recent work on HI and CO surveys. A slightly older review by Cox (2005) describes the distribution and morphological complexity of the ionized components and explores various unified models of ISM structure on the scale of the Galaxy. McKee et al. (2015) provide a comprehensive assessment of the stellar, ISM, and dark matter distribution in the solar neighborhood. The parameters of the superbubble shown in Figure 10.8 come from an analysis by McClure-Griffiths et al. (2003). Galactic surveys of HI and CO have been published by HI4PI Collaboration et al. (2016) and Dame et al. (2001) respectively, and include links to the data.

Questions

1. Show that points A and B in Figure 10.1 have the same V_{rad}.

2a. Write down an expression for the emission measure, EM, and dispersion measure, DM, for an ionized medium with uniform electron density, n_e, as observed along the line of sight to a pulsar at a known distance, d. Derive two expressions for n_e.

2b. In practice the two observations do not give the same value for the electron density. Show how the discrepancy can be reconciled if we allow the ionized medium to be clumped with a filling fraction ϕ, and derive expressions for n_e and ϕ.

3. Make back-of-the-envelope estimates for the total number of stars in a cluster with the size estimated to produce the superbubble in Figure 10.8. What is the expected distribution of OB stellar masses and main sequence lifetimes?

4. What is the thermal speed of the HIM? Make a simple argument (i.e., no detailed calculation) to explain why it is bound to the Galaxy.

Chapter 11
The ISM in Other Galaxies and Beyond

In this final chapter, we look at the ISM in other galaxies near and far. For certain aspects such as studying how spiral arms affect the gas, the lower spatial resolution and sensitivity due to the great distances are mitigated by the more favorable geometry. We can also explore a wider range of environments than exist in the Galaxy, such as galaxy mergers, starbursts, and low metallicities. Finally, a substantial fraction of the baryons in the Universe lie in the gas between galaxies and we can use the diagnostic tools that we developed for the ISM to learn about this intergalactic medium.

11.1 Nearby Galaxies

11.1.1 A Multi-Wavelength Panorama

Just as we began our description of the ISM with an overview of the Galaxy across the electromagnetic spectrum, a good starting point to learn about other galaxies is to look at them at different wavelengths. Figure 11.1 shows a montage of nine images, from the ultraviolet to the radio, of the spiral galaxy M51.

The face-on perspective allows us to see with great clarity the two-armed "grand design" pattern in stars, dust, and gas. The top row shows the stellar components from hot OB stars in the near-ultraviolet (NUV) to cool stars with spectral types K and M in the near-infrared. The HII regions created by the OB stars shine brightly through the Hα recombination line and bremsstrahlung radio emission. The cool dust associated with the denser parts of the ISM is seen in the mid- and far-infrared maps. The HI map shows velocity-integrated 21 cm line emission from atomic gas. The bottom right panel shows the distribution of molecular

Fig. 11.1. The M51 galaxy from ultraviolet to radio wavelengths. The top row with near-ultraviolet GALEX, Digitized Sky Survey red, and 2MASS K-band observations shows the stellar population from massive and hot to low mass and cool stars. In the second row, the Hα HST image and VLA 3 cm radio continuum map predominantly locate HII regions. The mid-infrared Spitzer and far-infrared Herschel maps show cool dust emission at ∼10−100 K. The last two panels of the bottom row show a VLA map of the HI 21 cm line from atomic gas, and an IRAM map of CO 1−0 emission from molecular clouds. Each panel is 9′ × 12′ in extent corresponding to 21 kpc×28 kpc at its distance of 8 Mpc.

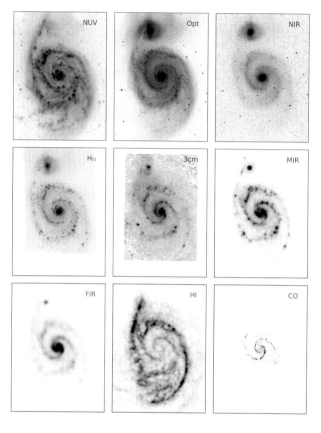

clouds via the velocity-integrated CO $J = 1−0$ line emission. This map is limited in size due to a combination of the relatively weak emission, atmospheric transmission, and the interferometric techniques required for high resolution at millimeter wavelengths.

These maps provide an overview of the relation between the dense components of the ISM, the stars that they form, and the ionized gas that they in turn produce. The atomic and molecular column-density profiles derived from the HI and CO maps are plotted in Figure 11.2 and can be compared to a similar plot for our Galaxy in Figure 10.3. The HI profile is quite similar in showing an exponential tail and a flattening in the inner region at $\sim 10^{25}$ m^{-2}. The H$_2$ profile also declines in a similar exponential fashion but the column densities are much greater than the HI and the values in our Galaxy. The total masses are $M_{HI} = 2.4 \times 10^9\ M_\odot$ and $M_{H_2} = 1.5 \times 10^{10}\ M_\odot$ showing that the ISM in M51 is predominantly molecular. Related to this, the ultraviolet data show many young stars that imply a total star formation rate, $\sim 15\ M_\odot$ yr^{-1}, five times greater than that of the Galaxy.

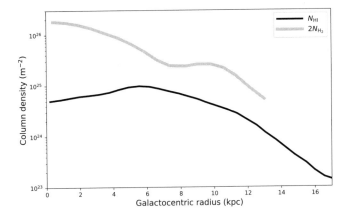

Fig. 11.2. Radial profiles of the surface density of atomic and molecular hydrogen gas across M51, in units of total number of hydrogen nuclei per square meter.

The flattening of the HI profile in the inner $\sim 10\,\mathrm{kpc}$ is also seen in other galaxies and occurs at a column density $\sim 10^{25}\,\mathrm{m}^{-2}$, equivalent to a mass surface density of $9\,M_\odot\,\mathrm{pc}^{-2}$. Above this level, the atomic gas efficiently converts to molecular. The center of M51 is noticeably deficient in HI. The CO emission is most prominent in the narrow spiral arms but the azimuthally averaged column density is highest in the center and the ratio of molecular to atomic gas reaches a peak of 40. As evidenced by the warm dust, HII regions, and stellar maps, the star formation rate is also very high in this region.

The molecular fraction differs substantially from one spiral galaxy to another and even more so when considering other galaxy types. To quantify the variation, however, requires understanding how to convert from the observed CO line intensity to an H_2 column density.

11.1.2 The CO-to-H_2 Conversion Factor

The invisibility of H_2 in cold molecular gas necessitates that we use some other tracer as a proxy. The low-energy rotational transitions of CO are the strongest lines from the molecular ISM that are readily detectable from the ground. However, a comparison with its isotopologues shows that the emission is optically thick. For instance, the peak CO 1–0 line brightness temperature is typically only about five times brighter than that of ^{13}CO 1–0 although their abundances differ by a factor of ~ 70 (as in star-forming regions in the Galaxy; see Figure 7.5).

Column densities of molecular regions in the Galaxy are most often now determined through observations of ^{13}CO or the even rarer $C^{18}O$, but this remains impractical for most extragalactic observations because the line strengths are so much weaker. When the only available data is a CO line, an approximate conversion to H_2 column density and thence mass is made using an empirical conversion factor,

$$X_{CO} = \frac{N_{H_2}}{W_{CO}}, \qquad (11.1)$$

where $W_{CO} = \int T_B dv$ is the velocity-integrated CO line intensity.

This use of this **X-factor** dates back to the early days of Galactic CO observations when instrumentation lacked the sensitivity to map the isotopologue lines. It was realized then that the Gaussian CO line shapes were inconsistent with optically thick emission from a uniform medium. Unresolved structures with a turbulent velocity field more readily explain the observations. Effectively, the emission is a sum of surfaces from numerous individual optically thick and unresolved structures. If higher column densities correlate with a greater range of motions, then the line broadens and the integrated line intensity correspondingly increases. This is schematically illustrated in Figure 11.3.

We can examine the sensitivity of X_{CO} to the properties of the ISM by a simple argument: the integrated intensity, $W_{CO} \propto T \Delta v$, and the column density, $N_{H_2} \propto M/R^2$, where the cloud temperature, linewidth, mass, and radius are $T, \Delta v, M,$ and R respectively. If the cloud is gravitationally bound, $M \propto R \Delta v^2$, then it follows that

$$X_{CO} \propto \frac{M/R^2}{T(M/R)^{1/2}} \propto \frac{\rho^{1/2}}{T}, \qquad (11.2)$$

where ρ is the average density. Molecular clouds in the solar neighborhood have similar temperatures and densities and therefore we would expect them to have similar X_{CO} conversion factors. Moreover, on cloud-size scales and larger, higher densities typically correspond with more active star formation and consequently higher temperatures. These

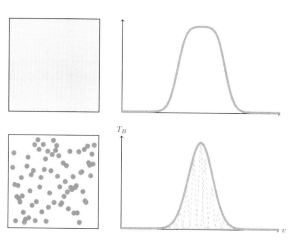

Fig. 11.3. The CO spectral line profile from a uniform, optically thick medium saturates and produces a flat-topped, non-Gaussian shape (top row). A collection of unresolved structures moving in a turbulent velocity field effectively contributes a number of narrow spectral features that sum up to produce a more Gaussian appearance (bottom row).

effects partially mitigate each other rendering X_{CO} relatively insensitive to environment.

The condition that the clouds be bound can be relaxed, leading to a more complicated set of dependencies but ending with a similar conclusion that, for the typical observed macroscopic properties of the molecular ISM, the CO integrated line intensity is indeed a useful proxy of the H_2 column density. Its value has been calibrated for Galactic clouds through cloud mass measurements via dust absorption and emission, optically thin molecular lines, as well as gamma-ray absorption from hydrogen nuclei. For the usual radio nomenclature where line strengths are expressed in terms of the brightness temperature, the Galactic value is

$$X_{CO} = 2 \times 10^{24}\,\mathrm{m}^{-2}(\mathrm{K\,km\,s}^{-1})^{-1}, \tag{11.3}$$

with an uncertainty for any given, uncalibrated object of $\pm 30\%$. Molecular regions in M51 and other spiral galaxies are estimated to have the same mean conversion factor but with a larger uncertainty $\gtrsim 50\%$.

In regions of lower metallicity, Z, there is less carbon and oxygen and therefore a lower $[CO]/[H_2]$ abundance. The effect on X_{CO} can be studied through PDR modeling (Chapter 7) or observations of Galactic clouds at large radii and other galaxies. For small variations, $X_{CO} \propto 1/Z$ but it then increases sharply for $Z \lesssim 0.3\,Z_\odot$. At such low metallicities in molecular regions, the CO column densities are too low for self-shielding to be effective.

11.1.3 Spiral Arms

The arms of spiral galaxies are most prominent in the ultraviolet and optical light that is produced by hot, luminous stars, the emission lines from their accompanying HII regions, and the dust and gas associated with their formation. Observing nearby galaxies in these various tracers shows the lifecycle of molecular clouds.

Spiral arms are not physically coherent structures in a galaxy but rather density waves that rotate with a radially independent pattern speed. They are generally trailing in the sense that the tails of the arms point away from the direction of rotation and have a logarithmic form, where the radius increases with azimuthal angle as $R \propto e^{b\theta}$ and b parameterizes the tightness of the spirals.

Galaxy rotation curves are approximately flat so the angular speed decreases inversely with radius. In the inner regions, gas will move into the relatively slow-moving spiral density pattern where it is shocked and compressed to high density. This results in rapid conversion of atomic gas to molecular and then gravitational collapse to star formation.

Fig. 11.4. Close-up of a spiral arm in M51 as seen in Hα with CO emission overlaid in contours. The gray line shows a logarithmic spiral and the three black arrows indicate circular rotations where the angular speed is faster for smaller radii.

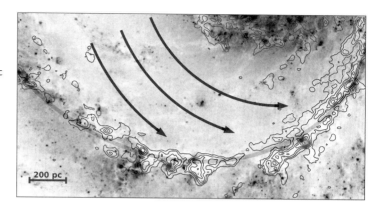

A close-up of an M51 spiral arm in Hα and CO is shown in Figure 11.4. The arrows indicate circular rotation representing the motions of ISM material into the arm. The compact knots of Hα emission are HII regions that signpost recently formed massive stars and lie slightly downstream from the molecular clouds. From the physical separation and rotation speed, the timescale between cloud formation and HII region breakout is estimated to be ~7 Myr.

The CO emission is clumpy, showing giant molecular clouds with masses $\sim 10^5 - 10^7\, M_\odot$. The high contrast between the arm and inter-arm regions shows that these objects both form and disperse rapidly with a lifetime $\sim 20 - 30$ Myr. The atomic gas is more broadly distributed and does not show a clear offset with the Hα map. The arm–interarm contrast is also markedly lower than for the CO. Most of the 21 cm emission is from atomic gas produced by photodissociation of the molecular clouds as they leave the spiral arm. The perpetual cycle through which the ISM transitions from diffuse to molecular and then back to diffuse as it rotates through the spiral arms converts about 5% of its mass into stars on each rotation.

The companion galaxy that is clearly seen in the optical and infrared images of Figure 11.1 drives the strong two-armed spiral in M51 that produces these unusually molecular rich regions and high star formation rate. Spiral galaxies are diverse in terms of number of arms, molecular to atomic ratio, star formation rate and efficiency. These structures are generally created by instabilities due to self-gravity of the disk, as quantified by the Toomre Q parameter (Equation 9.48).

11.1.4 Galaxy Collisions

M51 is a particular case where the spiral arms are strongly driven by the companion galaxy visible at the top of Figure 11.1. Galaxies form

804 nm

10 kpc

Fig. 11.5. HST I band image of an interacting galaxy pair known as "The Mice". The long stellar streams are stars that have been tidally stripped during previous encounters. Much of the ISM collided and remains near the center where new star formation takes place in the shocked, dense gas. Credit: NASA, Holland Ford (JHU), the ACS Science Team and ESA.

in clusters and are relatively close with typical separations of a few Mpc or $\sim$100 times the size of their stellar and gaseous disks (and $\sim$10 times the size of their dark matter halos). Consequently galaxy encounters are common, though they are generally not as neat as for M51, and play a major role in galaxy evolution. This is, again, a huge subject and we only cover an ISM-specific aspect here.

One of numerous possible examples of a galaxy collision is shown in Figure 11.5. Because of their large sizes, one side of the galaxy feels a significantly stronger gravitational force than the other which results in stretching and the long tidal tails. The gas behaves very differently from the stars however, because it has a much higher cross-section.

This is readily illustrated by a basic back-of-the-envelope estimate, where we consider two spiral galaxies comparable to our own passing through one another. The average distance between stars in each galaxy is about 1 pc so the stellar number density $n_* \sim 1\,\mathrm{pc}^{-3}$, and each star has a geometric cross-section $\sigma_* \sim \pi R_\odot^2$. The mean free path of a star during the galaxy collision is therefore $l = 1/(n_* \sigma_*) \sim 10^{15}\,\mathrm{pc}$. Compared to the scale of a galaxy, $L = 10\,\mathrm{kpc}$, the chance that a star would collide with another is $l/L \sim 10^{-11}$. As there are about 10^{11} stars in a galaxy we might expect about one head-on stellar collision. All the other stars would pass through unscathed though their motions would be affected by the larger scale gravitational potential.

From the ISM perspective, however, the encounter can be seen as two layers of gas about 30 kpc long and 3 kpc thick crashing into each other at the infall speed, which we can estimate is comparable to orbital motions, $v = 200\,\mathrm{km\,s}^{-1}$. Referring back to Figure 8.4, this produces a thin adiabatic region where the gas is heated and moderately compressed, followed by a broader zone where the gas cools and becomes much denser. The heating initially ionizes the gas but it recombines and becomes molecular in the post-shocked region on a timescale much shorter than the crossing time, $t_{\text{cross}} = L/v \simeq 50\,\text{Myr}$ (a handy conversion to use here is $1\,\text{km s}^{-1} \simeq 1\,\text{pc/Myr}$). Thus, well before the stellar encounter is over, the compressed clouds in their high-pressure environment undergo prolific star formation.

In addition, much of the ordered orbital angular momentum may be lost to an extent that depends on the geometry of the encounter. Gas can then fall toward the center of the gravitational potential where it can fuel additional star formation or the growth of a central AGN. In short, galaxy collisions dramatically and rapidly alter the properties of the ISM. As the galaxies merge together over many crossing times, the gas is almost fully converted to stars or tidally stripped away, with the end result of a gas-poor elliptical galaxy.

11.2 Extragalactic Star Formation

11.2.1 Scaling Relations

The large-scale perspective afforded by extragalactic observations allows us to define empirical relations between the gas and star formation that help us understand how galaxies work. By looking at scales much larger than giant molecular clouds, we average over the varied spatial and temporal characteristics of individual star-forming regions and can study their collective properties.

Using maps similar to Figure 11.1, we can compare the distribution of HI measured in the 21 cm line, H_2 scaled from CO observations by the X_{CO} factor, and star formation rate determined from $H\alpha$, infrared, or radio continuum observations. The star formation rate is best determined from observations of massive stars as they dominate the luminosity. The HII regions produce copious $H\alpha$ and radio continuum emission that directly relates to the ionization rate and therefore stellar number. Because stars form in dusty environments, the infrared luminosity is also a good measure of the star formation rate. A mixture of theory and observations calibrated against each other reveals the following linear relations,

$$\frac{\text{SFR}}{M_{\odot}\,\text{yr}^{-1}} = \frac{L(\text{H}\alpha)}{1.9 \times 10^{34}\,\text{W}} = \frac{L(\text{IR})}{2.6 \times 10^{36}\,\text{W}} = \frac{L_{\nu}(1.4\,\text{GHz})}{1.6 \times 10^{21}\,\text{W Hz}^{-1}}. \tag{11.4}$$

Not surprisingly, galaxies with greater masses of gas have higher overall star formation rates. Partly, this is simply a matter of scale so, to learn about the underlying physical processes, we normalize by area and look at the relation between surface densities,

$$\Sigma_{\text{SFR}} = A\Sigma_{\text{gas}}^{N}, \tag{11.5}$$

where Σ_{SFR} has units of $M_\odot \, yr^{-1} \, kpc^{-2}$, and Σ_{gas} is in $M_\odot \, pc^{-2}$. This power law form is known as the **Kennicutt–Schmidt law**. With well-resolved maps, we can extend this to comparisons across radial annuli within a galaxy or on a pixel-by-pixel basis down to kiloparsec scales.

The correlation of star formation with all the neutral ISM, $\Sigma_{gas} = \Sigma_{HI} + \Sigma_{H_2}$, gives a non-linear scaling, $N \sim 1.5$–2.5, with considerable scatter in the proportionality constant, A. However, the relation is much tighter and has a linear form, $N \approx 1$, when restricted to only molecular gas, $\Sigma_{gas} = \Sigma_{H_2}$. This shows that molecular clouds are the fundamental star-forming unit and the efficiency with which they convert to stars is constant,

$$\frac{\Sigma_{SFR}}{\Sigma_{H_2}} \approx 5 \times 10^{-10} \, yr^{-1}. \tag{11.6}$$

This means that 1% of the molecular mass is converted to stars every 20 Myr, which recalls the result in Chapter 9 that molecular clouds do not convert all their mass to stars on a free-fall timescale. The inverse of this ratio implies a timescale of 2 Gyr for the depletion of molecular gas. The similarity of this value from the centers to outskirts of an individual galaxy and from one galaxy to another, dwarf or spiral, indicates that star formation proceeds in a fairly regular way that depends only on the molecular reservoir.

Then what is the role of HI? Figures 10.3 and 11.2 show that the atomic disk extends much further than the stellar and molecular disks. When the gas density drops below a threshold of about $3 \, M_\odot \, pc^{-2}$, there is insufficient shielding of dissociating radiation, little molecular gas, and therefore no star formation. The radially averaged HI and H_2 column densities increase toward the center, but the H_2 grows faster than the HI and is tracked by the star formation rate. The HI column density saturates at $\sim 9 \, M_\odot \, pc^{-2}$ where the conversion to molecular form is efficient and leads to the highest star formation rates. Thus, the atomic gas is a fairly uniform background reservoir with a narrow range of column densities, only a factor of about 3, within the star-forming disk of a galaxy.

Star formation and galaxy evolution ultimately depend on the processes through which atomic gas converts to molecular and then back again as stars energize their surroundings. This interplay occurs below the kiloparsec scales where the Kennicutt–Schmidt law applies and brings together much of the ISM physics described in this book. In its most reductionist form, we envisage an atomic medium at a constant pressure set by the overlying weight of the gas (where the weight is determined by the gravitational field mainly from the stellar component in the disk). If the WNM cools, it transitions to the denser CNM, which is the essential prerequisite to the star-forming molecular gas. Stars then form, dissociate the molecules, heat the atomic gas, and induce the CNM to transition back to the WNM. Stellar feedback limits the formation efficiency from molecular gas and the overall cycle self-regulates through the balance between WNM/CNM fraction and heating.

In the early Universe, as galaxies accumulated most of their stellar mass, a more dynamical model may apply. Young disk galaxies at redshifts $z \sim 1 - 2$ ($\sim 3 - 6$ Gyr after the Big Bang) have higher molecular gas fractions, higher star formation rates, and are more turbulent than their counterparts today. The turbulence is greater than can be accounted for by stellar energy input and is attributed to gas infall from the intergalactic medium (see below) and gravitational instabilities in a dense disk where the Toomre $Q \lesssim 1$. These instabilities drive radial flows, concentrate gas, and create the conditions for extreme star formation events.

11.2.2 Super Star Clusters

The brightest, densest clusters contain thousands of OB stars forming in a single region. In the earliest stages, these are apparent as ultra-compact HII regions, similar to those discussed in Chapter 6 but dramatically scaled up in luminosity such that they can be detected in other, typically starburst, galaxies at Mpc distances.

As we will shortly see, starburst galaxies have radio SEDs that are dominated by synchrotron radiation from supernova remnants with a negative spectral index, $\alpha = d\log F_\nu/d\log \nu \simeq -0.5$. In some cases, however, high-resolution observations reveal parsec-scale knots of thermal emission with positive indices, $\alpha \approx 2$. These are bright, small, optically thick HII regions.

At wavelengths of about a centimeter and shorter, the emission flattens as the bremsstrahlung emission becomes optically thin (see Equation 6.21). For a turnover frequency $\nu_1 = 20$ GHz, the implied emission measures are extremely high, EM $\simeq 10^{21}$ m^{-6} pc. Their resolved sizes, $R \approx 1$ pc, then imply average densities for the ionized gas $\langle n_e^2 \rangle^{1/2} \approx 2 \times 10^{10}$ m^{-3}. This is similar to Galactic ultra-compact HII regions but with an order of magnitude larger size, corresponding to a thousand times larger volume and proportionally increased ionizing luminosity,

$$\dot{N}_{\text{ionize}} = \frac{4\pi}{3}\alpha_2 n_e^2 R^3 \simeq 10^{52}\, \text{s}^{-1}. \qquad (11.7)$$

This is a prodigious amount, orders of magnitude greater than most Galactic HII regions and equivalent to the output of $\sim 10^3$ O stars. There are huge numbers of associated lower mass stars of course. If a Salpeter IMF applies, the total stellar mass of the cluster is an astonishing $\sim 10^5\, M_\odot$.

555 nm

2 pc

Fig. 11.6. The R136 cluster in the Large Magellanic Cloud, as imaged in the optical by the Hubble Space Telescope. This is a young super star cluster, emerging from its ultra-compact state when the emission is dominated by the radio and infrared, on its way to becoming a globular cluster. Credit: NASA, ESA, F. Paresce (INAF-IASF, Bologna, Italy), R. O'Connell (University of Virginia, Charlottesville), and the Wide Field Camera 3 Science Oversight Committee.

The escape speed from such a massive, compact object, $v = (2GM/R)^{1/2} \simeq 30 \, \mathrm{km \, s^{-1}}$, is greater than the thermal linewidth of ionized gas at 10^4 K. Remarkably, the HII regions are gravitationally bound, which likely explains why they can form such a large number of energetic stars in close proximity to one another. Nevertheless, stellar winds eventually impart sufficient mechanical energy into the region that the gas ultimately breaks free and the cluster expands. Chapter 9 shows that the stellar cluster will remain bound if the star formation efficiency is greater than 50%. If this is indeed the case, these young super star clusters are the likely progenitors of globular clusters.

One of the closest examples of a super star cluster is R136 in the Large Magellanic Cloud (Figure 11.6). Though it still produces a bright HII region, it is no longer in the ultra-compact stage and has broken out and become optically visible. There are several extremely massive stars with $> 100 \, M_\odot$ here and the overall cluster has a radius of 2 pc with age 1.5 Myr and total stellar mass of $5 \times 10^4 \, M_\odot$.

Cluster R136 has already reached its final stellar mass. The ultra-compact phase when stars are actively forming probably only lasts for a few dynamical times before the HII region breaks free, $\sim 10^5$ yr, which implies a local star formation rate $\sim 1 \, M_\odot \, \mathrm{yr^{-1}}$. That is, over this short timescale, this single concentrated region creates almost as many new stars as our entire Galaxy does. Such extreme events require extreme conditions to form, namely huge quantities of gas rapidly concentrated into a small region. This can occur through massive cloud collisions in isolated cases in an irregular galaxy such as the Large Magellanic Cloud or on galactic scales through a merger.

11.2.3 Starburst Galaxies

Our Galaxy produces about $2\,M_\odot$ of stars per year whereas the rate in molecular-rich M51 is five times greater. Starburst galaxies have vastly greater star formation rates $\gtrsim 100\,M_\odot\,\mathrm{yr}^{-1}$. This rate cannot be produced by the compression of the ISM in spiral arms alone but instead is caused by large-scale conversion of diffuse to dense gas in galaxy collisions. The ISM is efficiently converted to stars but used up very quickly.

The molecular Kennicutt–Schmidt law, which relates the star formation rate density with the H_2 surface density, deviates from a linear scaling for starburst galaxies,

$$\Sigma_{\mathrm{SFR}} \propto \Sigma_{H_2}^{1.5}. \tag{11.8}$$

However, the relation reverts back to a linear form when only the dense molecular component observed in HCN or other tracers (see Table 7.1) is used,

$$\Sigma_{\mathrm{SFR}} \propto \Sigma_{\mathrm{dense}}. \tag{11.9}$$

This is analogous to the difference between counting all the neutral gas or just the molecular component in normal galaxies, and suggests that the fundamental unit that produces the rapid burst of star formation is the dense gas and that a non-linear process that converts low density H_2 to high density is the regulating factor.

Due to the obscuring layers of warm dust, most of the star-forming luminosity is emitted at infrared wavelengths, though there is an optical component as massive stars break free of their surroundings as in R136. The starburst occurs over tens to hundreds of Myr as the galaxies interact and is therefore accompanied by supernovae from the death of short-lived massive stars that form early on. This introduces an additional synchrotron radiation component. The SED of the closest starburst galaxy, M82, is shown in Figure 11.7. Note that this is on the scale of the entire galaxy, so localized regions such as the super star clusters with bright bremsstrahlung emission discussed above are a minor component here.

The SED is shown in terms of energy density, λF_λ, with a logarithmic wavelength scale so the energy output at each decade in wavelength can be readily compared with each other, $\lambda F_\lambda \Delta \log \lambda \simeq F_\lambda \Delta \lambda$. The dust emission peaks at $\sim 100\,\mu\mathrm{m}$ corresponding to a temperature $\sim 30\,\mathrm{K}$. Related silicate and PAH emission features are prominent in the mid-infrared. The steep falloff at longer wavelengths, $F_\lambda \propto \lambda^{-6}$, is a

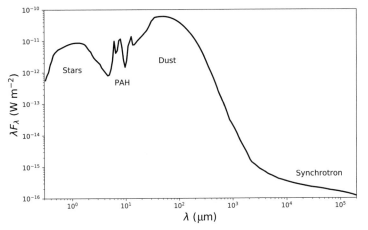

Fig. 11.7. The SED of starburst galaxy M82 plotted in terms of energy density, λF_λ. The plot spans over five orders of magnitude in wavelength from optical starlight to radio synchrotron emission powered by supernovae.

combination of the Rayleigh–Jeans decline of thermal emission and low emitting efficiency of small grains at long wavelengths.

Far-infrared surveys, first with IRAS and most recently with Herschel, show a plethora of high-redshift galaxies with similar SED shapes, demonstrating that merger powered starbursts are an essential component of galaxy evolution and the star-forming history of the Universe. As we look at different redshifts, the SED shifts to longer wavelengths and down due to distance. When viewed in terms of the observed quantity, flux density, this has the useful feature that the SEDs overlap at millimeter wavelengths, as demonstrated in Figure 11.8.

For observations at a fixed wavelength, say 1 mm, we therefore measure the emission from higher up the steep climb in the dust emission as the redshift increases. This almost exactly compensates for the dilution of the radiation due to the greater distance, with the result that the flux density remains approximately constant. Thus millimeter-

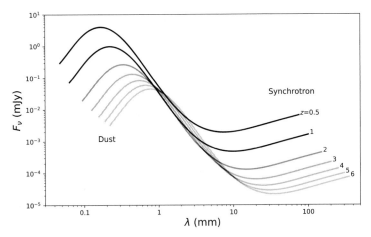

Fig. 11.8. The infrared–radio region of M82 plotted in terms of flux density, F_ν, as would be seen at redshifts from $z = 0.5$ to 6 (for a flat, ΛCDM cosmology with $H_0 = 67\,\mathrm{km\,s^{-1}\,Mpc^{-1}}$, $\Omega_M = 0.3, \Omega_\Lambda = 0.7$). The convergence of the curves in the central region shows that the flux density is roughly independent of redshift at millimeter wavelengths.

wavelength observations can uniquely survey the warm dust from star formation across the Universe with no redshift bias. Far-infrared cooling lines such as CI also shift into this band at high redshift but because they are at a fixed wavelength, their flux density decreases as the inverse square of the luminosity distance.

11.2.4 The First Stars

Stars form through the collapse of dense gas clouds where the gravity overcomes thermal pressure. In today's galaxy, this occurs in cold molecular clouds where rotational lines of CO and other molecules provide escape routes for radiative losses down to ~ 10 K. However, metals are formed by stellar nucleosynthesis, so the very first stars in the early Universe must have formed in different conditions.

The starting point is primordial neutral hydrogen clouds, with a 10% admixture of helium. As outlined in Chapter 5, Lyman α emission can cool the gas until collisions become too weak to excite the transition at $\sim 10^4$ K. At these temperatures, some H_2 can survive if it can form. In today's Galaxy, H_2 forms on dust grains but, due to the lack of metals, this process also cannot apply in the early Universe. Rather, a different catalytic process happens that relies on the free electrons left over from the recombination era,

$$H + e^- \rightarrow H^-,$$

$$H^- + H \rightarrow H_2 + e^-. \tag{11.10}$$

This two-body process is enhanced where the primordial density fluctuations are higher. Then the H_2 quadrupole rotational transitions, primarily $J = 2-0$ at 28.2 μm, provide the critical pathway to lower thermal pressures, higher densities, and ultimately runaway gravitational collapse.

Once the $[H_2]/[H]$ abundance reaches $\sim 10^{-3}$, the gas can cool down to ~ 200 K on a timescale shorter than the dynamical time at densities $n_H = 10^{10}$ m^{-3}. These temperature and density values translate to a Jeans mass of 350 $M_\odot$ (Equation 9.1), a characteristic value where gravity overcomes thermal pressure. The exact mass of the first stars is a much more complex problem as it requires understanding additional details in the formation process such as fragmentation and radiative feedback. Nevertheless, these ideas suggest that the first zero-metallicity, so-called **Population III**, stars were generally very high mass and may have formed within about 100 Myr after the Big Bang.

Such massive stars would have very short lives of at most a few Myr. They would then either collapse into black holes or explode as supernovae. If the latter, they would enrich their surroundings with metals that would seed a chain of accelerating star formation. These stars are now long gone of course, but searches for the signatures that their accompanying HII regions might have imprinted on the CMB are ongoing. Finally, the discovery of extremely low metallicity stars in the Galaxy, with $Z < 10^{-6}$ in the notation of Chapter 10, indicates that at least a few low-mass stars did form out of almost pure hydrogen gas.

11.3 The Intergalactic Medium

Just as the space between the stars in the Galaxy is not completely empty, so it is with the space between galaxies. Cosmology tells us that about 5% of the matter in the Universe is baryonic but galaxy number counts show that only about 10% of this is in stars. The remaining 90% is in the intergalactic medium (IGM). The critical density for a flat Universe is

$$\rho_{crit} = \frac{3H_0^2}{8\pi G} = 8.5 \times^{-27} \text{ kg m}^{-3}, \tag{11.11}$$

where $H_0 = 67 \text{ km s}^{-1} \text{ Mpc}^{-1}$ is the Hubble constant. This implies an average IGM hydrogen number density $n_H \simeq 0.2 \text{ m}^{-3}$, over six orders of magnitude lower than the ISM.

The timescales associated with such a low-density gas are extremely long, whether for collisional or radiative interactions. It is therefore extremely hard to observe except in its primordial state as neutral hydrogen in the early Universe.

11.3.1 The Lyman α Forest

The first atoms in the Universe formed once the Universe had expanded sufficiently to cool down to about 3000 K. Until the first stars formed, the Universe consisted almost entirely of neutral hydrogen and helium. Nevertheless, within a gigayear, supermassive black holes had formed producing luminous quasi-stellar objects (QSOs) which act as beacons for us to detect the atomic gas and watch it become ionized as the Universe continued to expand and was flooded with radiation.

The accretion of gas onto a black hole efficiently converts gravitational energy to radiation via viscous heating. The extremely hot, fast-moving gas produces a thermal spectrum with substantial ultraviolet photons and a broad Lyman α line at $\lambda = 121.6$ nm produced by recombination of newly ionized gas. If unimpeded, these photons would stream out through the expanding Universe and reach us as a faint continuum and redshifted Lyman α line in the optical. However, any neutral hydrogen along the way would strongly absorb the continuum that is

Fig. 11.9. Optical spectrum of a QSO at redshift $z = 4.520$. The bump at the far right is redshifted Lyman α emission from the QSO and the shorter wavelengths to the left show a forest of Lyman α absorption lines due to neutral hydrogen at intervening redshifts along the line of sight. The inset shows the continuum normalized absorption from a damped Lyman α system. The observations were taken with the ESI spectrograph on the Keck telescope on Maunakea, Hawaii.

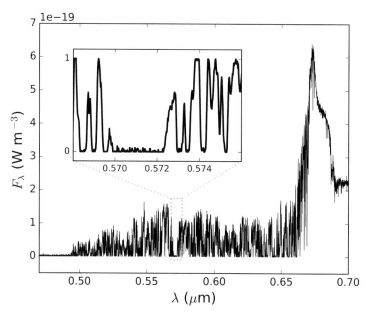

redshifted to 121.6 nm in its local rest frame. The excited atoms would quickly re-emit but in all directions so effectively scattering the light out of our line of sight. The result is that the continuum shortward of the QSO Lyman α line shows a series of hydrogen absorption features.

For high-redshift QSOs such as shown in Figure 11.9, the density of absorption lines is so great that it is termed a Lyman α forest. The wavelength of each absorption line translates to the redshift and therefore distance of each absorber, and the equivalent width relates to its column density through the curve of growth techniques described in Chapter 5. Most lines have low optical depths with column densities $N_{\rm HI} < 10^{21} \, {\rm m}^{-2}$. However, there are a few very broad, deep features known as damped Lyman α systems. This is the effect of the Lorentz line wings at extremely high optical depths, $\tau > 10^3$, and column densities, $N_{\rm HI} > 10^{24} \, {\rm m}^{-2}$. These dense regions are associated with the gas bound to galaxies, known as their **circumgalactic medium**. The inset in Figure 11.9 shows the most prominent example in the spectrum.

From observations of QSOs along different lines of sight and over different wavelength ranges, extending to Hubble Space Telescope ultraviolet observations for low-redshift objects, we can map out the evolution in the mass and morphology of the neutral IGM. Most of the hydrogen mass lies in the damped Lyman α systems, which are primordial galaxies or a circumgalactic medium. A small fraction of lines can be identified through their redshifted frequencies as metals, from which we can learn about enrichment of young galaxies and the IGM.

Even with the forest of lines, the fact that the continuum can be outlined in Figure 11.9 shows that much of the IGM is ionized. Beyond $z \sim 6$, however, QSOs show almost complete absorption at redshifts shortward of the Lyman α line due to multiple, overlapping lines. This is known as a Gunn–Peterson trough and implies a high fraction of neutral gas in the IGM. The evolution is very rapid around this redshift indicating a phase change in the IGM ionization state, similar to the boundary of HII regions in Chapter 6. The race is on to detect the faint HI hyperfine structure line, redshifted to ~ 1.5 m, from this epoch.

11.3.2 The Nearby IGM

The Lyman α forest becomes increasingly bare as we look at lower redshift QSOs implying a decreasing amount of neutral hydrogen. The IGM is almost completely ionized by $z = 2$ corresponding to a lookback time of 10 Gyr or about 3 Gyr after the first stars. Even with the $(1 + z)^3$ volume scale factor, the densities are so low that the recombination time, $1/\alpha_2 n_e \sim 20$ Gyr, is greater than a Hubble time and therefore, once the gas is ionized, it will forever remain so.

The same reasoning shows that all two-body interactions are vanishingly rare and radiative losses are minimal. The gas therefore heats up, mainly via shocks, to temperatures $\gg 10^5$ K. As with the HIM in the Galaxy, such hot and diffuse gas is hard to detect. However, the ultra-low densities in the IGM are compensated by the great distances and the nearby IGM can be detected through absorption lines of OVI and OVII against bright X-ray emitting AGN known as blazars. A promising new avenue for continued exploration of the IGM is through studies of recently discovered fast radio bursts. Though generally irregular or even singular events, these strong extragalactic flares act as analogs of pulsars in the Galaxy and the dispersion of their broad radio emission (Chapter 6) allows us to infer the IGM electron density along the line of sight. With polarization measurements of Faraday rotation, the presence and magnitude of an intergalactic magnetic field might also be determined, but that is a topic beyond the scope of this book as we have reached the end of our journey.

Notes

The multi-wavelength montage of M51 in Figure 11.1 comes from data at dustpedia.astro.noa.gr (uv, optical, infrared), archive.stsci.edu/prepds/ m51 (Hα), www.mpifr-bonn.mpg.de/3168319/m51 (3 cm), www.mpia .de/THINGS (HI), and www2.mpia-hd.mpg.de/PAWS/PAWS (CO). Of particular note, the dustpedia website has multi-wavelength data for many galaxies and the PAWS website links to many papers that discuss the CO data in relation to the stellar and other ISM components. The parameters of the radio observations of super star clusters come from the discovery paper of this class of objects in Turner et al. (2000). The QSO spectrum in Figure 11.9 was obtained from a public release of the data associated with Prochaska et al. (2007). More details on the development of a telescope in South Africa designed to detect the HI 21 cm line in the early Universe can be found at reionization.org.

Questions

1. Restate Equation 11.1 in terms of a relation between CO luminosity, $L_{CO} = \int W_{CO} dA$, and molecular mass, where the constant of proportionality has units of $M_\odot\,(\mathrm{K\,km\,s^{-1}})^{-1}\,\mathrm{pc}^{-2}$.

2. Consider the super star cluster described in Subsection 11.2.2 in a galaxy at a distance of 5 Mpc. What is the angular size and flux density at the turnover frequency? What telescope could be used to observe this?

3. An excellent primer for cosmological distance formulae is Hogg (1999). Using the luminosity distance for the same cosmological parameters as Figure 11.8, create a series of redshifted blackbodies with a temperature of 30 K. Is the flux density at 1 mm approximately constant? Compare with the M82 SED and explain why the inefficient emission of dust grains at long wavelengths allows us to see starburst galaxies across the Universe.

Appendix
Constants in SI and cgs Units

Table A.1. Physical constants

	SI	cgs
Speed of light, c	$3.00 \times 10^8 \, \mathrm{m \, s^{-1}}$	$3.00 \times 10^{10} \, \mathrm{cm \, s^{-1}}$
Planck constant, h	$6.63 \times 10^{-34} \, \mathrm{J \, s}$	$6.63 \times 10^{-27} \, \mathrm{erg \, s}$
Boltzmann constant, k	$1.38 \times 10^{-23} \, \mathrm{J \, K^{-1}}$	$1.38 \times 10^{-16} \, \mathrm{erg \, K^{-1}}$
Stefan–Boltzmann constant, σ	$5.67 \times 10^{-8} \, \mathrm{W \, m^{-2} \, K^{-4}}$	$5.67 \times 10^{-5} \, \mathrm{erg \, s^{-1} \, cm^{-2} \, K^{-4}}$
Gravitational constant, G	$6.67 \times 10^{-11} \, \mathrm{m^3 \, kg^{-1} \, s^{-2}}$	$6.67 \times 10^{-8} \, \mathrm{cm^3 \, g^{-1} \, s^{-2}}$
Hydrogen mass, m_H	$1.67 \times 10^{-27} \, \mathrm{kg}$	$1.67 \times 10^{-24} \, \mathrm{g}$

Table A.2. Astronomical constants

	SI	cgs
Parsec, pc	$3.09 \times 10^{16} \, \mathrm{m}$	$3.09 \times 10^{18} \, \mathrm{cm}$
Astronomical unit, au	$1.50 \times 10^{11} \, \mathrm{m}$	$1.50 \times 10^{13} \, \mathrm{cm}$
Solar mass, $M_\odot$	$1.99 \times 10^{30} \, \mathrm{kg}$	$1.99 \times 10^{33} \, \mathrm{g}$
Solar luminosity, $L_\odot$	$3.83 \times 10^{26} \, \mathrm{W}$	$3.83 \times 10^{33} \, \mathrm{erg \, s^{-1}}$
Solar radius, $R_\odot$	$6.96 \times 10^8 \, \mathrm{m}$	$6.96 \times 10^{10} \, \mathrm{cm}$

Glossary

absorption coefficient The proportion of specific intensity removed per unit length along the line of sight as radiation passes through a region.

accretion disk A disk where material is falling onto the central object.

accretion luminosity The luminosity from the gravitational energy released by the accretion of matter onto a central object.

adiabatic The case where changes in a system occur without loss of energy.

angular resolution The angular scale on which distinct objects can be distinguished.

Alfvèn speed The propagation speed of waves in a magnetized fluid.

blackbody radiation Radiation with a specific intensity equal to the Planck function.

bolometric luminosity The power emitted by a source summed over all frequencies and wavelengths.

bolometry A process to characterize incident radiation by measuring the change in the temperature of the detector.

Boltzmann distribution The distribution of quantized energy levels of particles in a system.

Bonnor–Ebert mass Maximum stable mass of a hydrostatically supported core.

brightness temperature Radio specific term related to the specific intensity.

central molecular zone The region within about 200 pc around the Galactic Center that contains high-density, warm molecular gas.

Chandrasekhar–Fermi method A relation between the magnetic field strength with the ratio of turbulent line width to polarization angle dispersion.

charge-coupled device Electronic imager consisting of an array of photodetectors that are read out sequentially.

circumgalactic medium The gas that is bound to a galaxy beyond its stellar component and any rotationally supported disk.

cold neutral medium Relatively dense and cold neutral atomic gas in the ISM, with temperatures of $\sim 100\,\mathrm{K}$.

collision strength Dimensionless scaling factor in the expression for the collisional rate coefficient of a particle.

collisional rate coefficient The volumetric rate for collisional de-excitation.

color The magnitude difference of a source between two specified wavelengths.

column density The number of particles per unit area along the line of sight.

continuum Radiation that is observed over a wide range of wavelengths.

cosmic rays Relativistic particles (protons and heavier atomic nuclei) produced by interstellar shocks.

cosmochemistry The study of the chemical composition of objects in the Universe.

critical density The density at which the collisional excitation rate is balanced by radiative decay.

curve of growth The increase in equivalent width with peak optical depth of a line.

dark cloud A dense region of the ISM that blocks most optical light from background stars.

detailed balance The principle which states that every elementary process is statistically balanced by its exact reverse process when a system is in equilibrium.

dipole moment A measurement of the polarity, related to the separation between the center of charge and center of mass, in a molecule.

dispersion measure The integral along the line of sight of the electron density.

dispersion relation A relation between the spatial and temporal frequencies of a wave.

Doppler effect The process by which the frequency of radiation is shifted by the motion of an object toward or away from the observer.

emission coefficient The radiative energy produced by a source in units of specific intensity per unit length.

emission measure The integral along the line of sight of the square of the electron density.

equation of state The relation between the pressure, density, and temperature in a system.

equivalent width A measure of the area, in wavelength units, under the continuum of an absorption line.

excitation temperature A temperature defined from the distribution of energy levels using the Boltzmann distribution.

extinction The reduction, in magnitudes, of radiation due to its passage through the ISM.

extinction curve The variation of interstellar extinction with wavelength.

extinction efficiency The ratio of the extinction to geometric cross-section.

flux The power emitted by a source per unit area, summed over a range of frequencies or wavelengths.

flux density The power emitted by a source per unit area at a particular frequency or wavelength.

forbidden lines Spectral lines observed at low densities in the ISM that are not seen in typical terrestrial conditions due to low decay rates.

giant molecular cloud A massive molecular cloud, with a mass $\gtrsim 10^5 \, M_\odot$ extending over several tens of parsec.

grain surface reactions Chemical reactions that use dust grains as a catalyst.

grain An interstellar dust particle.

graybody A Planck radiation spectrum multiplied by a frequency-dependent opacity term.

Hayashi track The path in the Hertzsprung–Russell diagram that a low mass pre-main sequence star takes as it contracts toward the main sequence.

Herbig Ae/Be stars Intermediate mass, optically visible, pre-main sequence stars.

Hertzsprung–Russell diagram A plot of luminosity versus temperature (or spectral type), revealing the main sequence and evolutionary states of stars.

heterodyne detection A technique to combine a synthesized signal with an astronomical signal and thereby change its frequency without loss of information.

high-velocity clouds Atomic clouds that are falling onto the Galaxy rather than following it's rotation curve.

hot ionized medium Very hot and rarefied ionized gas in the ISM, produced by shocks.

infrared cirrus Diffuse infrared emission produced by warm dust.

initial mass function The distribution of stellar masses before they evolve off the main sequence.

integrated line intensity Integral of a spectral line, generally expressed as intensity times velocity.

interferometric array A set of telescopes that are linked to one another to produce a higher resolution image of an object.

interstellar shock The abrupt change of fluid properties over a small region in the ISM.

inverse P-Cygni profile A spectral profile that shows red-shifted self-absorption.

ion–molecule reaction The step in a chemical reaction pathway that transfers a proton from an ionized to a neutral species.

isothermal The case where changes in a system occur at constant temperature.

isotopologue A molecule that consists of at least one less abundant isotope of its constituent elements.

jansky A unit of flux density with SI units $W\,m^{-2}\,Hz^{-1}$.

Kelvin–Helmholtz timescale The characteristic timescale for a star to lose gravitational energy through radiation.

Kennicutt–Schmidt law A power law relationship between the gas surface density and star formation rate.

kinematic distance A distance determined from an observed velocity, typically assuming a form for orbital motions.

Kirchoff's law The statement that the efficiency with which an object emits light is equal to the efficiency with which it absorbs light at the same wavelength.

line profile The shape of a spectral line with respect to frequency or wavelength.

Local Bubble A localized region around the Sun of diffuse, partially ionized gas.

local thermodynamic equilibrium A situation in which thermodynamic equilibrium holds in a restricted region.

Mach number The speed of motion in units of the sound speed of the surrounding medium.

magnetohydrodynamics The study of fluid motions in the presence of magnetic fields.

Maxwell–Boltzmann distribution The distribution of particle speeds in a gas of a given temperature.

metals Elements heavier than hydrogen and helium in the ISM.

monochromatic luminosity The power emitted by a source at a particular frequency or wavelength.

optical depth The exponential scaling factor for the reduction in intensity as radiation passes through a region.

partition function Scale factor used to convert a measure of column density in a single transition to the total column density of the species.

photodiode A device that produces an electric current in response to incident light.

photodissociation region Alternative name for photon dominated region.

photon dominated region The interface between atomic and molecular gas where the physical and chemical conditions are largely dictated by the incident radiation field.

Planck function The specific intensity of thermal radiation from a source with a given temperature.

polycyclic aromatic hydrocarbons A large molecule consisting mainly of hydrogen and carbon.

Population III The first stars that formed in the Universe with zero metallicity.

pre-main sequence star Young star recently formed from a molecular core, generally optically visible and contracting to smaller radii and lower luminosities toward the main sequence.

protostellar cluster A group of protostars or pre-main sequence stars.

ro-vibrational spectrum The spectrum produced by transitions between a set of vibrational and rotational energy levels of a molecule.

rotation diagram A plot of the column density versus energy for different transitions of a molecule.

rotational constant Characteristic frequency that provides the scaling for the rotational energy levels of a diatomic molecule.

selection rules A set of formulae relating quantum energy levels with high transition probabilities between them.

selective extinction The ratio between the extinction at V band and B-V color, characterizing the form of the optical extinction law.

self-absorption The absorption of radiation by matter that is part of the same physical region that is emitting.

self-shielding The reduction in the rate of dissociation of a molecule due to the emission of radiation through bound energy levels.

singular isothermal sphere Analytic solution for a hydrostatically supported core that tends toward infinite density at its center.

sound speed Speed at which a pressure or density enhancement travels through gas.

specific intensity The radiative energy passing through a region, specified per unit time, per unit area, per unit frequency or wavelengfth, per solid angle.

spectral energy distribution The variation of continuum radiation over a wide range of wavelengths or frequencies.

spectral lines Radiation emitted or absorbed in a narrow range of frequencies or wavelengths.

spectral resolution The wavelength scale on which two narrow spectral features can be distinguished from one another.

star formation efficiency The efficiency, measured by mass ratio, with which a particular region forms stars.

star formation rate The rate at which stars are formed in a particular region.

Stokes number The product of the stopping time and rotation rate for a particle in a gaseous disk.

Stokes parameters A description of the degree and angle of polarization of light.

superbubble A very large region of diffuse hot gas produced by clusters of OB stars.

T Tauri stars Optically visible pre-main sequence stars, named after the first identified object of its type.

thermodynamic equilibrium The state of a system where there are no changes in its macroscopic properties.

Toomre criterion for stability A condition between the sound speed, epicyclic frequency, and surface density for a differentially rotating disk to be gravitationally stable.

triggered star formation A process by which newly formed stars can induce the formation of a subsequent generation of stars.

virialized A state of dynamical balance between kinetic and gravitational energies in a system.

viscosity A measure of the friction between two layers of a shearing disk.

warm ionized medium Diffuse ionized gas in the ISM, produced by the leakage of ionizing photons from HII regions.

warm neutral medium Warm, diffuse neutral atomic gas in the ISM with temperatures $\sim 10^4$ K.

X-factor Constant of proportionality between the H_2 column density and CO line intensity.

Zeeman effect Splitting of a spectral line due to a magnetic field.

References

Armitage, P. J., *Astrophysics of Planet Formation* (Cambridge University Press, 2009, Second Edition 2020)

Buether, H., Klessen, R., Dullemond, C., et al., *Protostars and Planets VI* (University of Arizona Press, 2014)

Carroll, B. W. and Ostlie, D. A., *An Introduction to Modern Astrophysics* (Addison-Wesley, 2006, reissued by Cambridge University Press, 2017)

Clarke, C. and Carswell, B., *Principles of Astrophysical Fluid Dynamics* (Cambridge University Press, 2014)

Cox, D. P., 2005, *The Three-Phase Interstellar Medium Revisited.* Annual Reviews of Astronomy and Astrophysics, 43, 337

Dame, T. M., Hartmann, D., and Thaddeus, P., 2001, *The Milky Way in Molecular Clouds: A New Complete CO Survey.* Astrophysical Journal, 547, 792

Draine, B. T., *Physics of the Interstellar and Intergalactic Medium* (Princeton University Press, 2011)

Field, G. B., Goldsmith, D. W., and Habing, H. J., 1969, *Cosmic-Ray Heating of the Interstellar Gas.* Astrophysical Journal Letters, 155, L149

Herbst, E. and van Dishoeck, E. F., 2009, *Complex Organic Interstellar Molecules.* Annual Reviews of Astronomy and Astrophysics, 47, 427

Heyer, M. and Dame, T. M., 2015, *Molecular Clouds in the Milky Way.* Annual Reviews of Astronomy and Astrophysics, 53, 583

HI4PI Collaboration, Ben Bekhti, N., Flöer, L., et al., 2016, *HI4PI: A Full-Sky H I Survey Based on EBHIS and GASS.* Astronomy and Astrophysics, 594, A116

Hildebrand, R. H., 1983, *The Determination of Cloud Masses and Dust Characteristics from Submillimetre Thermal Emission.* Quarterly Journal of the Royal Astronomical Society, 24, 267

Hogg, D. W., 1999, *Distance Measures in Cosmology.* arXiv e-prints, astro-ph/9905116

Kalberla, P. M. W. and Kerp, J., 2009, *The H*ɪ *Distribution of the Milky Way*. Annual Reviews of Astronomy and Astrophysics, 47, 27

Krumholz, M. R., *Star Formation* (World Scientific Publishing, 2017)

Lada, C. J., Alves, J. F., and Lombardi, M., *Near-Infrared Extinction and Molecular Cloud Structure*. In B. Reipurth, D. Jewitt, and K. Keil, editors, *Protostars and Planets V*, 3 (University of Arizona Press, 2007)

Licquia, T. C. and Newman, J. A., 2015, *Improved Estimates of the Milky Way's Stellar Mass and Star Formation Rate from Hierarchical Bayesian Meta-Analysis*. Astrophysical Journal, 806, 96

Martins, F., Schaerer, D., and Hillier, D. J., 2005, *A New Calibration of Stellar Parameters of Galactic O Stars*. Astronomy and Astrophysics, 436, 1049

Mathis, J. S., Rumpl, W., and Nordsieck, K. H., 1977, *The Size Distribution of Interstellar Grains*. Astrophysical Journal, 217, 425

McClure-Griffiths, N. M., Dickey, J. M., Gaensler, B. M., et al., 2003, *Loops, Drips, and Walls in the Galactic Chimney GSH 277+00+36*. Astrophysical Journal, 594, 833

McClure-Griffiths, N. M., Dickey, J. M., Gaensler, B. M., et al., 2005, *The Southern Galactic Plane Survey: H I Observations and Analysis*. Astrophysical Journal Supplement Series, 158, 178

McKee, C. F. and Ostriker, E. C., 2007, *Theory of Star Formation*. Annual Reviews of Astronomy and Astrophysics, 45, 565

McKee, C. F. and Ostriker, J. P., 1977, *A Theory of the Interstellar Medium: Three Components Regulated by Supernova Explosions in an Inhomogeneous Substrate*. Astrophysical Journal, 218, 148

McKee, C. F., Parravano, A., and Hollenbach, D. J., 2015, *Stars, Gas, and Dark Matter in the Solar Neighborhood*. Astrophysical Journal, 814, 13

Ossenkopf, V. and Henning, T., 1994, *Dust Opacities for Protostellar Cores*. Astronomy and Astrophysics, 291, 943

Osterbrock, D. E. and Ferland, G. J., *Astrophysics of Gaseous Nebulae and Active Galactic Nuclei* (University Science Books, 2006)

Parthasarathy, R., Frank, C., Treffers, R., et al., 1998, *A Rooftop Radio Observatory: An Undergraduate Telescope System at the University of California at Berkeley*. American Journal of Physics, 66, 768

Planck Collaboration, Aghanim, N., Alves, M. I. R., et al., 2016, *Planck Intermediate Results. XXXIV. The Magnetic Field Structure in the Rosette Nebula*. Astronomy and Astrophysics, 586, A137

Prochaska, J. X., Wolfe, A. M., Howk, J. C., et al., 2007, *The UCSD/Keck Damped Lyα Abundance Database: A Decade of High-Resolution Spectroscopy*. Astrophysical Journal Supplement Series, 171, 29

Rieke, G. H., *Measuring the Universe* (Cambridge University Press, 2017)

Rybicki, G. B. and Lightman, A. P., *Radiative Processes in Astrophysics* (Wiley-VCH, 1986)

Savage, B. D. and Sembach, K. R., 1996, *Interstellar Abundances from Absorption-Line Observations with the Hubble Space Telescope.* Annual Reviews of Astronomy and Astrophysics, 34, 279

Schlafly, E. F., Meisner, A. M., Stutz, A. M., et al., 2016, *The Optical-Infrared Extinction Curve and Its Variation in the Milky Way.* Astrophysical Journal, 821, 78

Tielens, A. G. G. M., *The Physics and Chemistry of the Interstellar Medium* (Cambridge University Press, 2010)

Turner, J. L., Beck, S. C., and Ho, P. T. P., 2000, *The Radio Supernebula in NGC 5253.* Astrophysical Journal Letters, 532, L109

Ward-Thompson, D. and Whitworth, A. P., *An Introduction to Star Formation* (Cambridge University Press, 2015)

Wolfire, M. G., McKee, C. F., Hollenbach, D., et al., 2003, *Neutral Atomic Phases of the Interstellar Medium in the Galaxy.* Astrophysical Journal, 587, 278

Wood, D. O. S. and Churchwell, E., 1989, *The Morphologies and Physical Properties of Ultracompact H II Regions.* Astrophysical Journal Supplement Series, 69, 831

Zucker, C., Speagle, J. S., Schlafly, E. F., et al., 2019, *A Large Catalog of Accurate Distances to Local Molecular Clouds: The Gaia DR2 Edition.* Astrophysical Journal, 879, 125

Index